Greenhouse Gas Management

Topical Conference at the 2010 AIChE Spring Meeting
and 6th Global Congress on Process Safety

San Antonio, Texas, USA
21 – 25 March 2010

ISBN: 978-1-61738-438-7

Printed from e-media with permission by:

Curran Associates, Inc.
57 Morehouse Lane
Red Hook, NY 12571

Some format issues inherent in the e-media version may also appear in this print version.

Printed by Curran Associates, Inc. (2010)

For permission requests, please contact AIChE
at the address below.

AIChE
3 Park Avenue
New York, NY 10016-5991

Phone: (203) 702-7660
Fax: (203) 775-5177

www.aiche.org

Additional copies of this publication are available from:

Curran Associates, Inc.
57 Morehouse Lane
Red Hook, NY 12571 USA
Phone: 845-758-0400
Fax: 845-758-2634
Email: curran@proceedings.com
Web: www.proceedings.com

Greenhouse Gas Management

Topical Conference at the 2010 AIChE Spring Meeting
and 6th Global Conference on Process Safety

San Antonio, Texas, USA
21 - 25 March 2010

ISBN: 978-1-61738-438-7

AIChE

TABLE OF CONTENTS

Impact of Carbon Cap and Trade on Refinery Energy Conservation Economics

Ian M. Glasgow, Stan Polcar, Earl Davis, Tram Nguyen, Raj Subbiah, Jerry Price, Chris Stuecheli, R.E. "Ed" Palmer
Mustang Engineering, L.P., Houston, TX

A program directed at the reduction of greenhouse gas (GHG) emissions is the object of increased focus in US public policy. This is evident by the House of Representatives passing *The American Energy and Security Act* (H.R. 2454) in June 2009 and similar legislation formerly proposed by Lieberman/Warner. The EPA is actively evaluating avenues in which the Clean Air Act can be used to reduce GHG emissions as a result of their determination that these substances are an endangerment to human health and welfare. In H.R. 2454 individual refiners would be responsible for GHG emissions from their manufacturing operations plus the emissions from the combustion of the fuels they sell. These total emissions represent about 35% of the total US GHG inventory; however in H.R. 2454 the refining sector is given 2% of the available emission allowances per year until 2025. This will result in refiners having to either purchase or find offsets for over 90% of their regulated GHG emissions.

In this potential carbon constrained economy there are likely to be significant incentives to implement energy conservation projects that are marginal or uneconomic based only on the value of fuel savings. In this paper several case studies will be presented where refinery energy conservation options will be considered with the economics for GHG emissions reductions included. A sensitivity analysis on the value of the GHG allowances will be included in the economic evaluation.

1. Introduction

GHG emissions have been an object of increased focus in US public policy. GHG are generally defined as the total of six compounds whose distribution for emissions in 2008 as reported by the Energy Information Administration (EIA) is shown in Figure 1. The main contribution is energy related CO_2 which is a result of fossil fuel combustion.

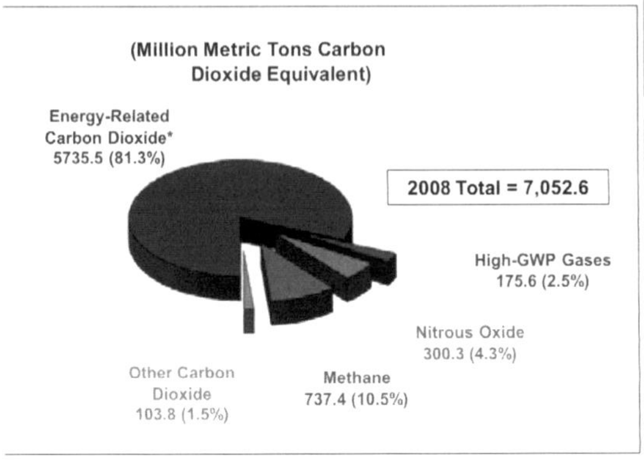

Figure 1: GHG Distribution (EIA, 2008)

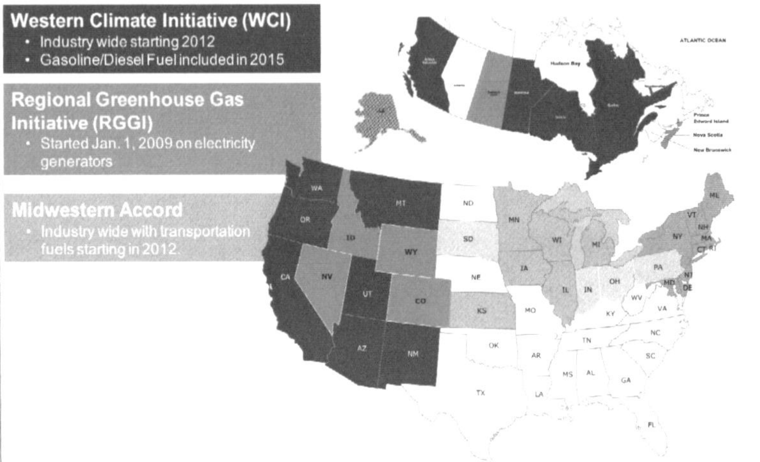

Figure 2: Regional Climate Initiatives

In June 2009 the US House passed the Waxman-Markey bill or *The American Clean Energy and Security Act*. The bill is a multi-tiered initiative with the centerpiece being a cap and trade system to reduce GHG emissions.

Of the GHG allowances available in the proposed cap and trade program each year, 2% will be given to the refining sector. This compares to about 35% of all US GHG that are actually emitted during the refinery manufacturing process and the subsequent combustion of fuels produced by various components of the transportation sector. As a result refiners will incur a significant expense to reduce manufacturing GHG emission and to purchase additional allowances required to comply with the cap and trade rules.

Other regulatory approaches are being evaluated by the EPA as a result of the final rulemaking associated with the endangerment finding of GHG to human health and welfare. This consists of mandatory reporting of GHG emissions, and a proposed rulemaking that stationary sources will require best available control technology (BACT) of GHG emissions for Title V and PSD Permits.

Previous enacted legislation that impact GHG emissions includes the 2007 Energy Independence Act Renewable Fuel Standard (RFS-2) and increased CAFÉ standards that were announced in June 2009. A comparison of RFS-2 and the recent revision to CAFE to the proposed Waxman-Markey bill shows that until 2020 these provisions will keep pace with reducing transportation vehicle GHG emissions as shown in Figure 3.

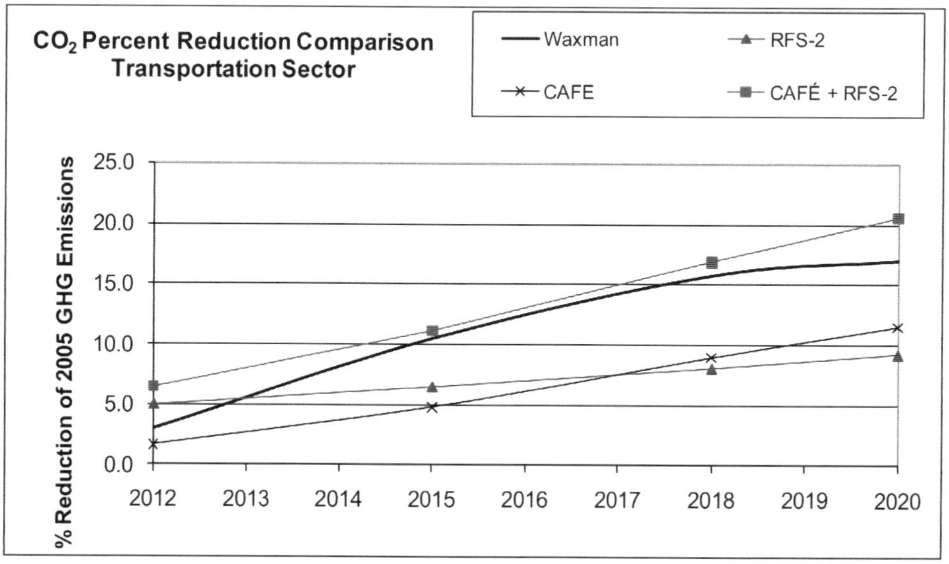

Figure 3: Comparison of Proposed Waxman-Markey to RFS-2 and CAFE.

Further in the absence of federal legislation regional cap and trade markets has been established as shown in Figure 2.

2. Case Studies

A 150 MBPSD notional refinery is used to demonstrate the magnitude of GHG emissions and potential energy efficiency projects. Arab Medium crude was selected as the basis. It is assumed that hydrogen is supplied from an onsite generation facility and a Catalytic Reformer (CCR). Gas oil upgrading is via a Cat Feed hydrotreater and FCC. Resid upgrading is via a Delayed Coker. It is also assumed that Flare Gas Recovery has already been installed. The notional refinery produces the following products:

- 8.4 MBPSD LPG
- 78.8 MBPSD Gasoline
- 14.8 MBPSD Jet Fuel
- 43.0 MBPSD ULSD
- 3.4 MBPSD No. 6 Fuel Oil
- 1,260 ton/day coke

GHG Emissions for the notional refinery are shown in Figure 4. Emissions from the combustion of transportation fuels sold by the refinery make up 80% of the potentially regulated emissions. Emissions that result from co-products (coke and LPG) are another 10%. Only 10% of the emissions are a result of the refinery manufacturing process.

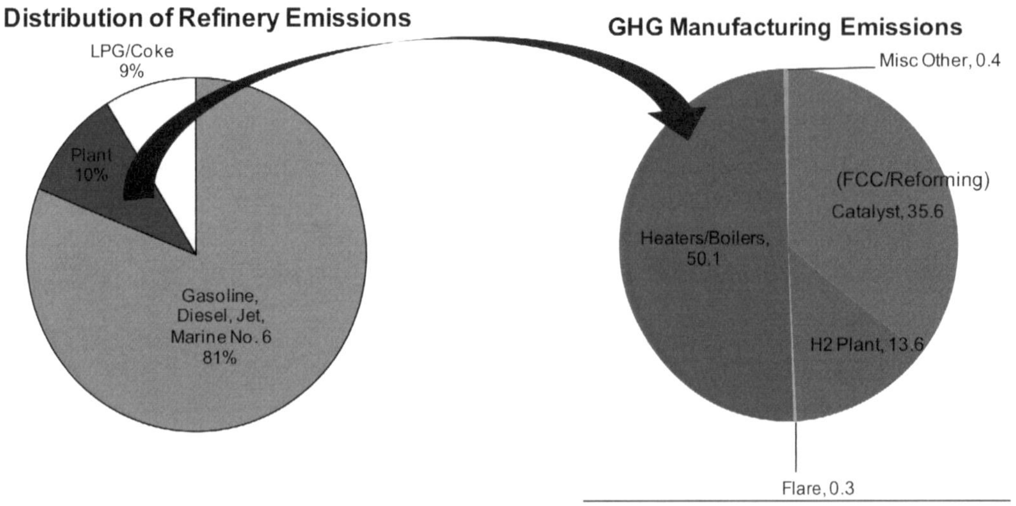

Figure 4: Notional refinery GHG emissions.

Incremental fuel for the refinery is assumed to be supplied by natural gas. It is assumed that GHG changes in electricity consumption related to offsets are compared to a coal fired power plant. The refinery is assumed to operate 8,400 hours per year. Energy prices are assumed to be $5.50/mmBTU for natural gas on a LHV basis, $7.70/Mlb of steam, and $0.07/kWh of electricity. The target Internal Rate of Return (IRR) for energy plant energy conservation projects is assumed to be 20%.

Following is a list of energy conservation projects that will be reviewed:

- Revamp Crude Preheat Exchange Train to increase the Crude Heater Inlet Temperature
- Add Combustion Air Preheat to Fired Heaters
- Replace the Vacuum Tower Steam Ejectors with a Liquid Ring Compressor
- Diesel HDS Hot Feed with additional Steam Generation
- Power Recovery from FCC Regenerator hot flue gas.

A review of the economics over a range of GHG emission prices is provided for each case.

2.1. Crude Heater Inlet Temperature

For this case it is assumed that the crude unit was originally sized for 110 MBPSD of feed and achieved a heater inlet temperature of 500°F. Over time the crude rate has increased to 150 MBPSD by minor revamps with no changes to the preheat exchange network. The existing preheat train uses four pumparounds from the Crude Tower (top, kero, diesel, AGO) and two pumparounds from the Vacuum Tower (LVGO and HVGO). All side cut products from the Crude and Vacuum tower are cooled against the crude feed including the vacuum tower bottoms (VTB). A summary flow sheet is provided in Figure 5.

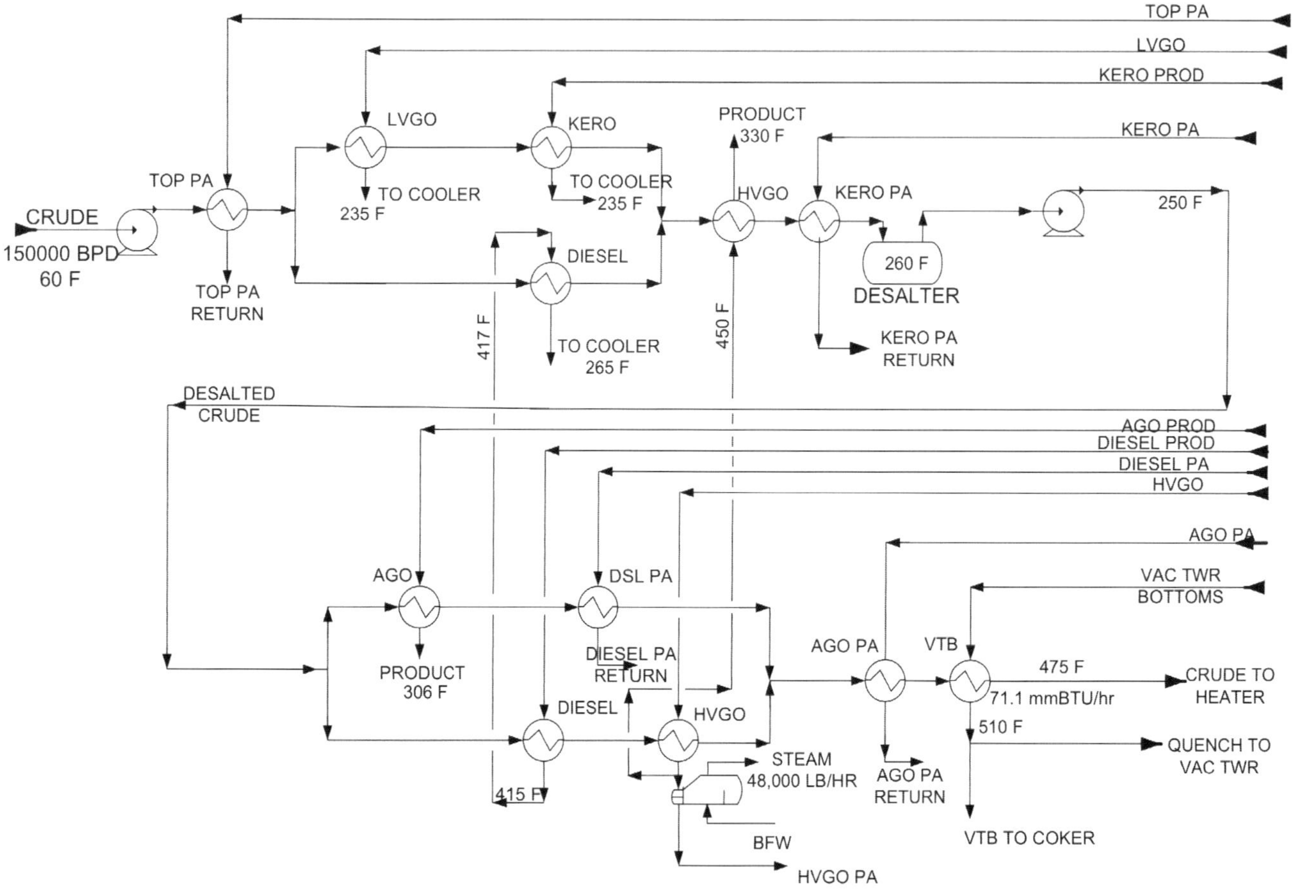

Figure 5: Current crude preheat exchange network of notional refinery.

It is the goal of this project to increase this crude heater inlet temperature to reduce the crude heater duty. The revamp design attempts to target thermal pinch limits to obtain maximum available duty using the existing hot streams that are available.

Care was taken to account for any change in waste heat steam production and the subsequent change in fuel consumption/GHG emissions at the boiler house.

Figure 6 shows a revised heat exchange configuration that result in the temperature to the charge heater increasing to 515°F. New exchangers were sized with appropriate metallurgy and it was assumed that existing plot space was available for the new equipment. Since a parallel path of exchangers were selected for the revamped crude preheat train, no changes were required for the pumps. The capital cost for this revamp was estimated to be $19.5 MM.

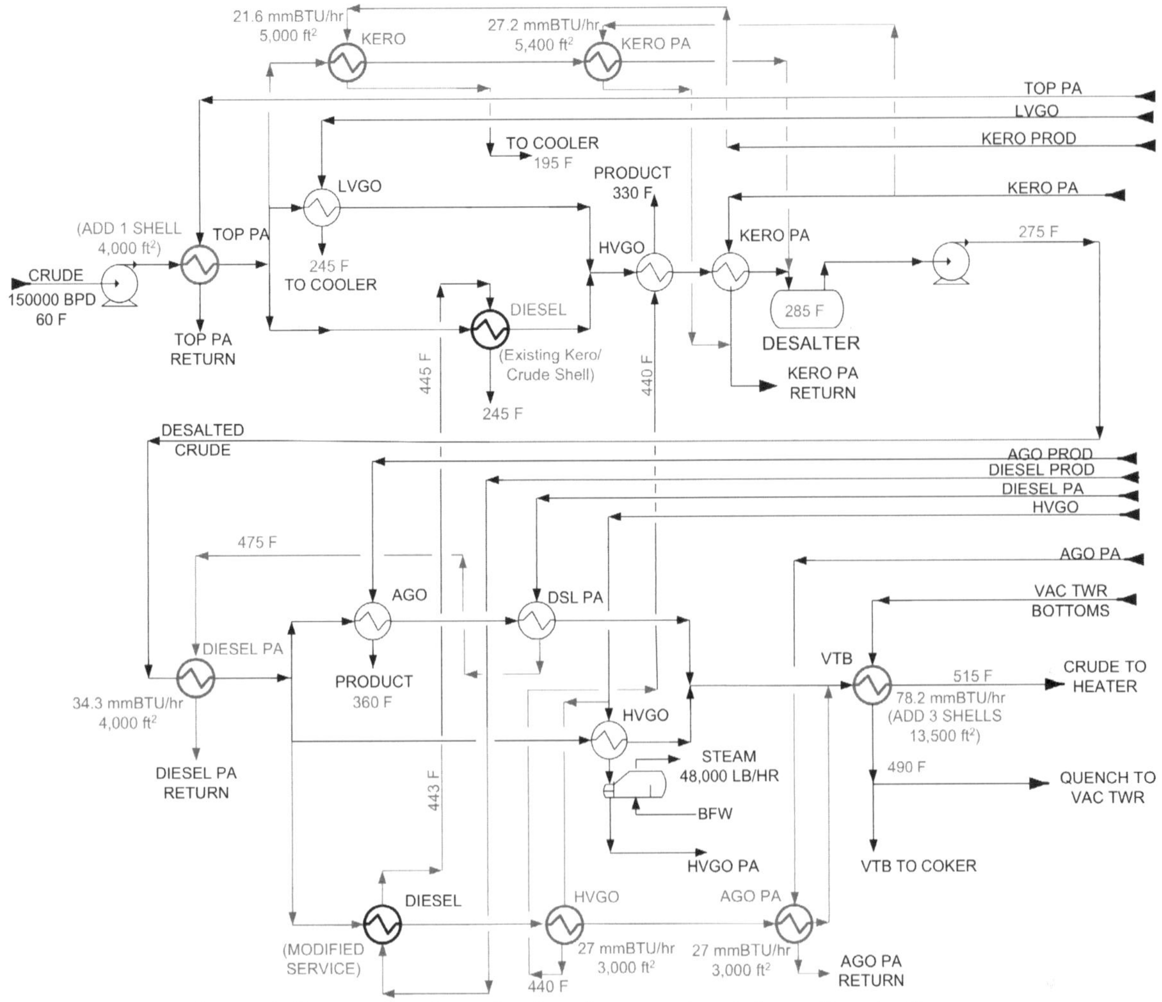

Figure 6: Revamp crude preheat exchange network for notional refinery.

The IRR for GHG emissions prices from \$0 to \$100/metric ton is summarized in Figure 7. A debit in steam generated in the Coker due to the decreased feed temperature from the vacuum tower is accounted for in the economics. The result shows that such a revamp is marginal based on fuel savings alone. However with the additional cost that will result from GHG emissions in a proposed cap and trade system the economics are significantly more attractive. A GHG cost of around \$50/metric ton is required to achieve the 20% IRR target.

Table 1: Summary of crude preheat exchanger network revamp results.

Heater		Base Case	Revamp Case
Absorbed Duty	mmBtu/hr	382.6	324.9
Steam in Convection Section	mmBtu/hr	36.1	28.3
Efficiency	%	87.0	87.4
Fired Duty	mmBtu/hr	481.25	404.1
Reductions			
Steam Generated from VTB	mmBtu/hr	0.0	-7.1
GHG Emissions	metric ton/yr	0	27,247

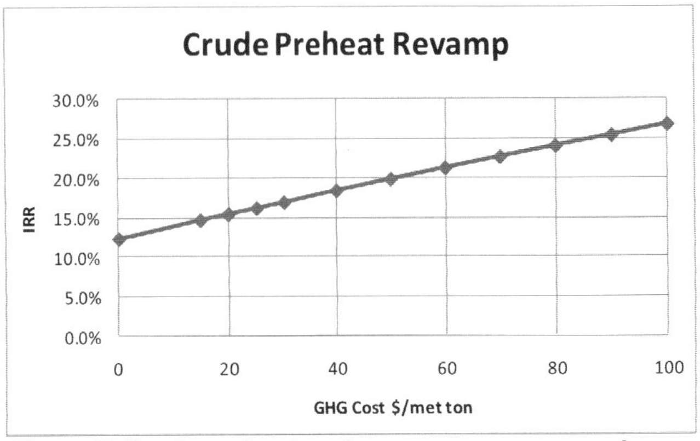

Figure 7: Crude preheat exchange revamp economics with cost of GHG emissions included.

An additional option that could achieve an even greater increase in crude heater inlet temperature by achieving tighter thermal pinch for targeted exchangers is the use of welded plate heat exchangers or other newer exchanger technologies.

2.2. Combustion Air Preheat

Traditionally, most refinery heaters have been a natural draft design with the process fluid flowing through both the convection and radiant sections. For larger heaters the additional convection surface may be used for steam generation to obtain efficiencies of 88 – 89 percent.

For large heaters where higher efficiencies can be justified, or if incremental steam generation capability has little or no value, combustion air preheaters have been used. Several past optimization studies have indicated that when fuel gas value is relatively high, air preheat is economically attractive for heaters above 75 to 80 mmBtu/hr absorbed duty.

Smaller heaters with older designs may have few convection rows with efficiencies of about 82 percent. With the additional cost of GHG emissions, these heaters may become good candidates for air preheaters. It is assumed the addition of air preheat will increase the heater efficiency from 82% to 92%.

Cost data for an air preheat systems equipment were obtained[1] as well as typical cost factors for installation with either revamped or new fired heaters[2]. The results for addition of an air preheat system to a new heater is shown in Figure 8.

The results show a significant reduced absorbed heater duty that is economically viable (20% IRR) as a result of cap and trade. At $40/metric ton air preheat will be viable for a grassroot heater with an absorbed duty just over 20mm Btu/hr. With GHG valued at $15/metric ton the threshold to meet the 20% IRR target is 40mmBtu/hr.

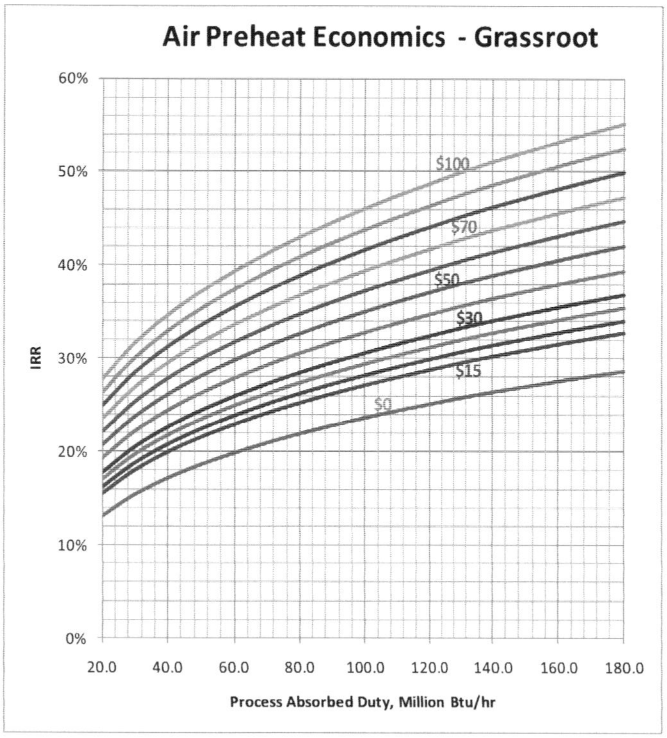

Figure 8: Grassroot heater air preheat economics with GHG emission costs included.

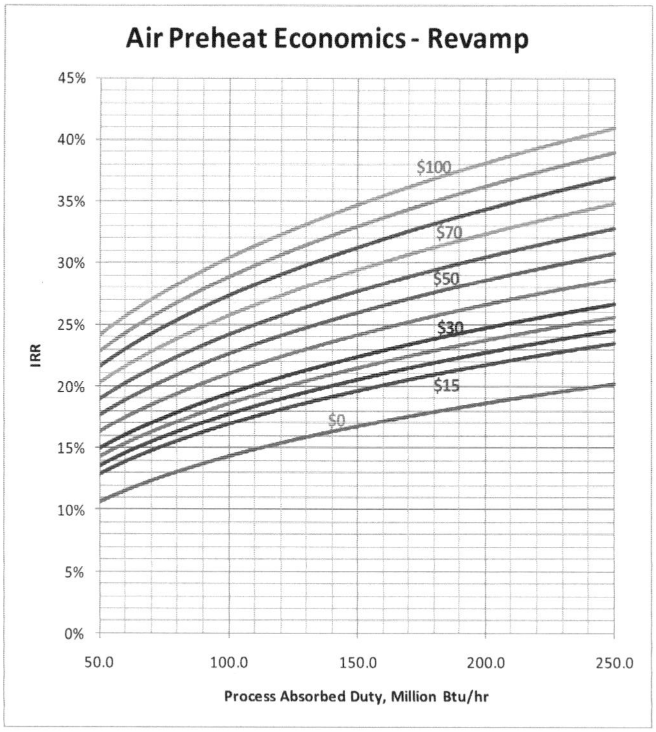

Figure 9: Revamp heater air preheat economics with GHG emissions cost included.

Figure 9 shows the economics for a revamp of an existing heater with air preheat. The TIC factor is larger for a revamp due to the amount of work required to the existing duct work, stack and other potential

1 – Data provided by OnQuest and Tulsa Heaters, Inc (THI)

2 – Data provided by Furnace Improvements

modifications. Larger heaters with an absorbed duty of just over 150 mmBtu/hr absorbed duty would meet the 20% IRR target with GHG emissions valued at $15/metric ton. A sensitivity analysis for natural gas prices at $6.50/mmBtu (versus $5.50/mmBtu for the base case) on a LHV basis showed benefits at $15/metric ton would extend to heaters with an absorbed duty down to 100 mmBtu/hr.

2.3. Vacuum Tower Steam Ejectors

The Crude Unit Vacuum Tower requires a reliable vacuum system to consistently maximize heavy vacuum gas-oil yields. The notional refinery has a three-stage ejector system as shown in Figure 10. Each ejector is coupled with an inter/after condenser with all condensed materials sent to a common seal drum. The motive fluid is 250 psig, 500°F MP steam and cooling water is available at 90 °F.

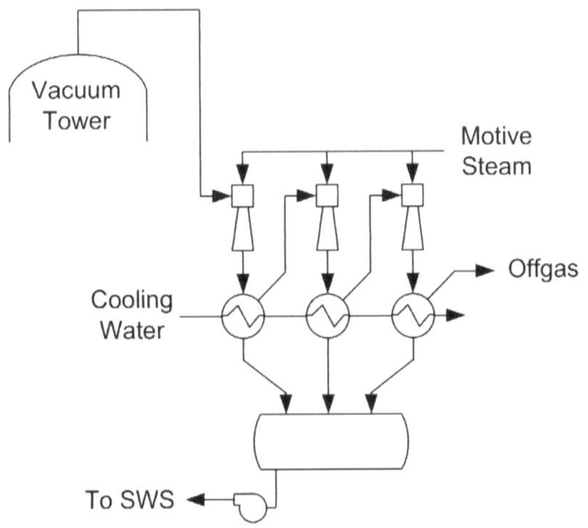

Figure 10: Current vaccum steam ejector system for notional refinery.

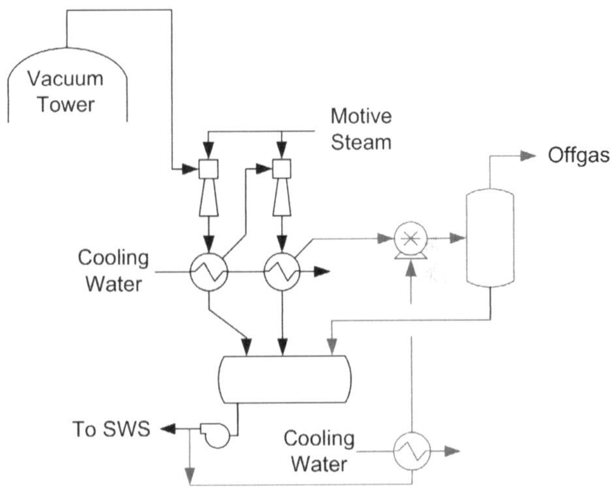

Figure 11: Vacuum steam ejector system revamp with liquid ring compressor.

Depending on the tradeoff between the costs of steam vs. electricity, it may be economical to replace an ejector stage with a liquid ring vacuum pump (LRVP). The LRVP option has lower steam usage which means reduced GHG emission at the boilers.

Graham Corporation was asked to provide a design basis for a hybrid system with associated equipment cost. The typical hybrid set-up has an LRVP as the last stage of a multiple-stage system. An LRVP is not technically feasible with the 1st stage steam ejector. Graham evaluated the replacement of the 2nd stage ejector with LRVP, but it was also not practical because of the high volumetric rate of the process gas. A summary of the process data comparing the two cases is provided in Table 2.

The retrofit scope leaves the 3rd stage ejector and condenser abandoned in place, while the LRVP skid is located at grade. This would allow the 3rd stage ejector to be used when maintenance is performed on the LRVP. Plot space is assumed to be available near the vacuum tower.

An installed capital cost of $2.8 MM estimate was developed using the equipment cost provided by Graham and installation cost determined by Mustang. The seal fluid configuration is closed-loop to minimize the seal water make-up rate. The typical skid from Graham includes seal fluid recirculation pumps, which were not needed by the notional refinery. Figure 11 shows the hybrid system.

The IRR for GHG emissions prices from $0 to $100/metric ton is shown in Figure 12. The result shows that such a revamp would be marginal on fuel savings alone. However with the additional cost that would result from GHG emissions in a proposed cap and trade program, the project has significantly improved economics.

Table 2: Summary of vacuum steam ejectors revamp results.

Demand		**Base Case**	**Hybrid Case**
Steam	lb/hr	32,152	21,473
Cooling Water	gpm	2,650	2,735
Electicity	bhp	0	217
Reductions			
Boiler Fired Duty	mmBtu/hr	0	14.9
GHG Emissions	metric ton/yr	0	6,649

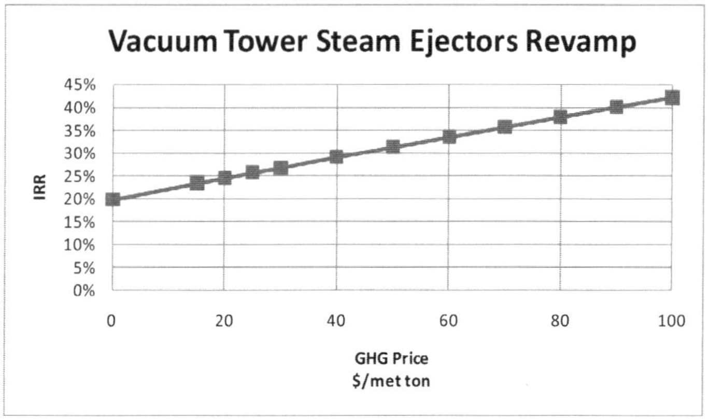

Figure 12: Vacuum Tower steam ejectors system revamp economic with GHG emissions cost included.

2.4. Unit Hot Feed and Steam Generation

The notional refinery for this study has a Diesel HDS unit that process cold diesel feed from storage to produce ULSD. Cold feed is preheated in a low pressure stripper bottoms exchanger. The reactor effluent heats the LP Separator liquid (Stripper feed) and generates steam prior to being collected in the downstream flash drums. It is assumed that this steam generation was economically justified during the conversion to ULSD. A schematic of the process is provided in Figure 13.

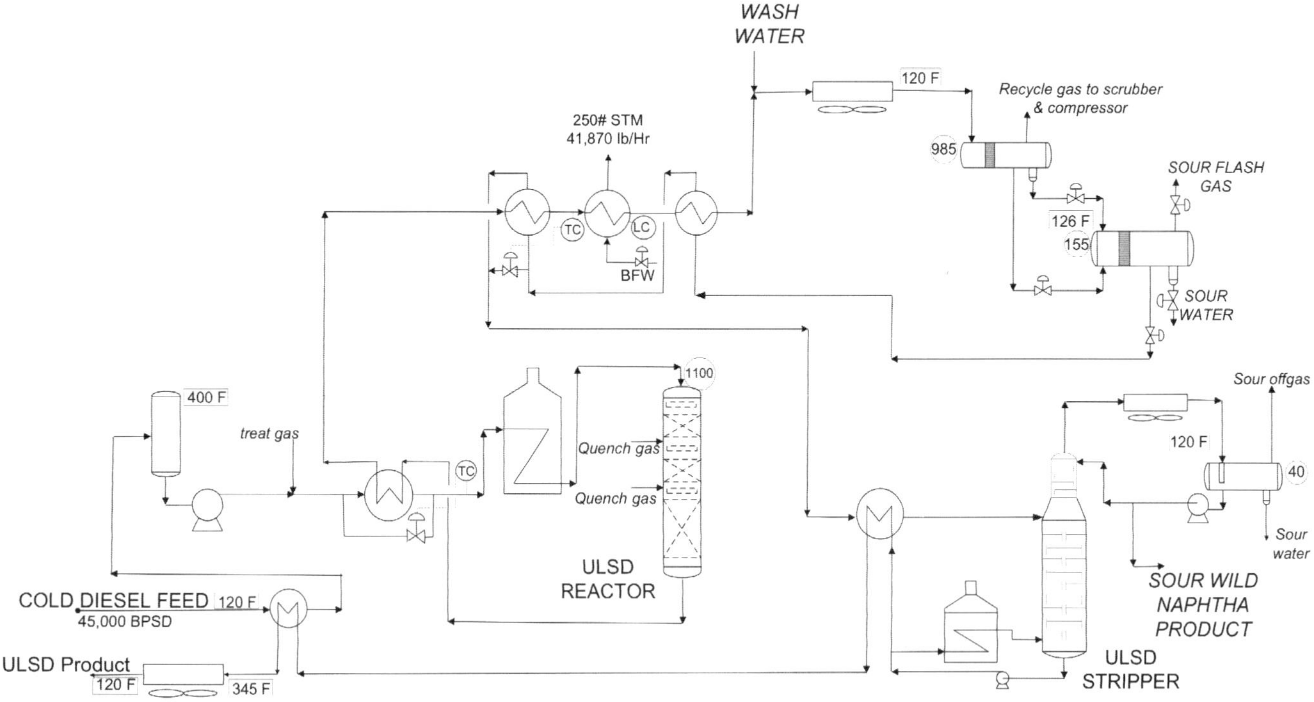

Figure 13: Current Diesel HDS system of notional refinery.

For this study, generation of additional 250 psig saturated MP steam was evaluated. A promising target for heat exchange was identified by utilizing a hot unit feed to reduce the unit feed/product exchanger duty. The additional cooling needed for product storage can be achieved via steam generation. A summary of the process data is provided in Table 3.

A cost estimate was developed for the process revamp. 1,000 ft of new piping was included for each new feed line. The estimated installed cost is $3.2 MM. The revamp process schematic is shown in Figure 14.

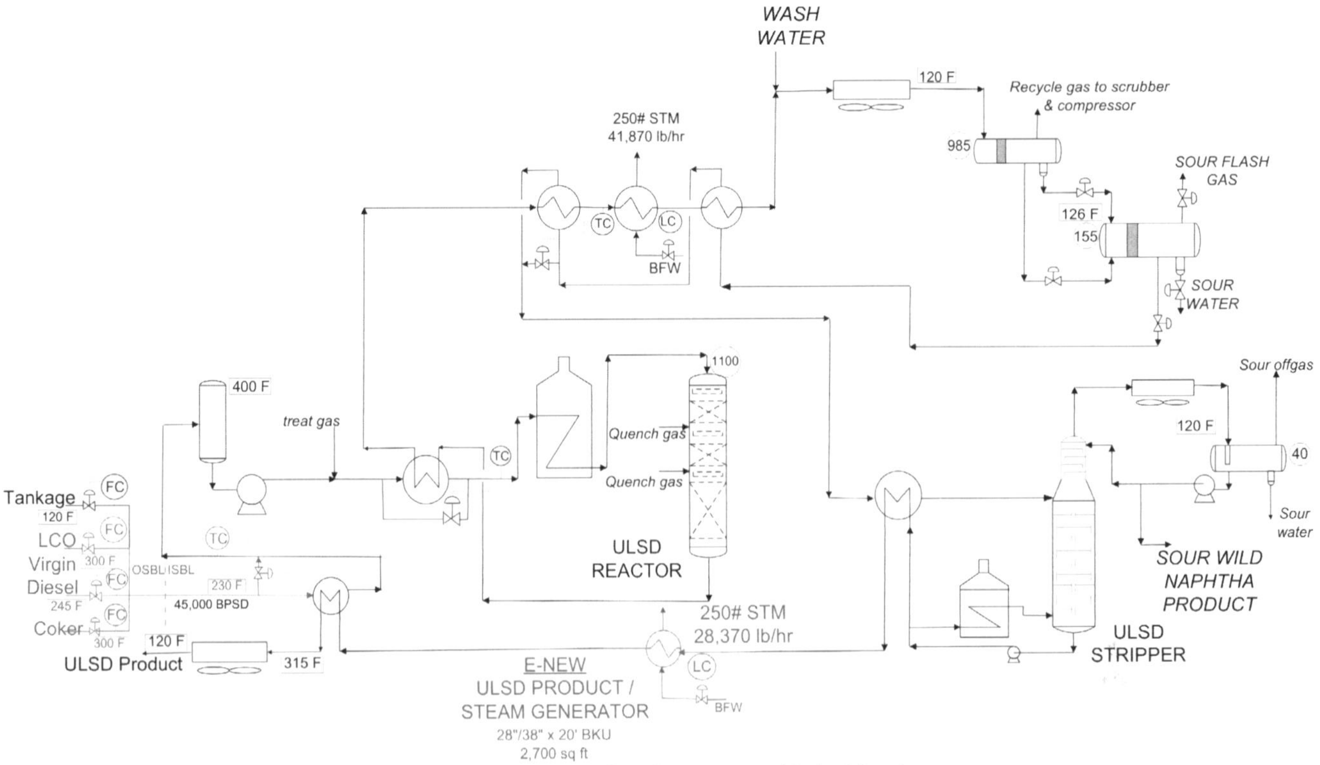

Figure 14: Diesel HDS unit revamp with hot feed.

The results show that the economics with current energy prices would make the project economically viable without GHG credits. Figure 15 shows that the economies are enhanced significantly when GHG emissions costs are considered.

Table 3: Summary of Diesel HDS unit hot feed revamp results.

Generated		**Base Case**	**Revamp Case**
Steam	lb/hr	41,870	69,370
Boiler Fired Duty	mmBtu/hr	46.6	79.1
Reductions			
GHG Emissions	metric ton/yr	0	14,489

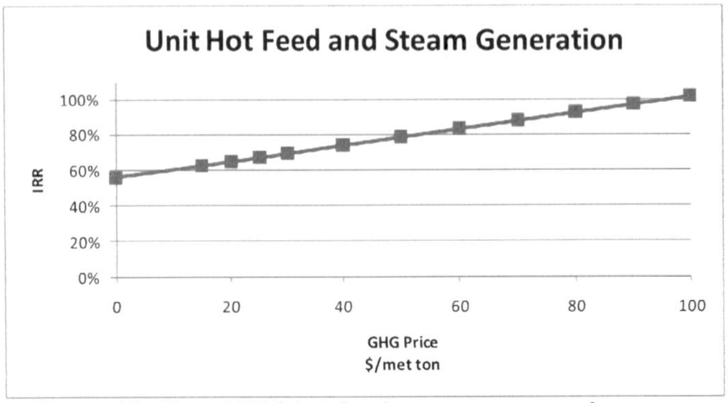

Figure 15: Diesel HDS hot feed revamp economics with GHG emissions included.

2.5. *Power Recovery*

The FCC unit has a high temperature flue gas stream from the Regenerator that is typically used to make high pressure steam. The flue gas is typically at a pressure of 30-35 psig with a temperature of around 1,350°F. The flue gas is then let down to 1.5 psig through an orifice before being routed to the flue gas cooler and the emission controls system (electrostatic precipitator or wet gas scrubber). In the flue gas cooler 600 psig, 600°F HP steam is generated using a steam generator with an economizer and superheat section. This process schematic for our notional refinery is shown in Figure 16.

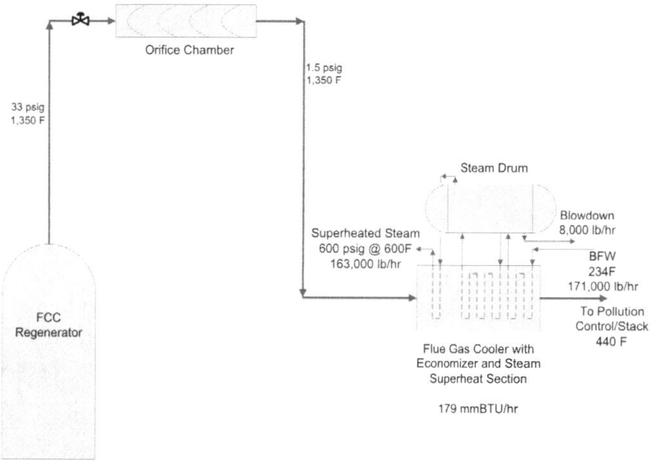

Figure 16: Existing FCC flue gas system of notional refinery.

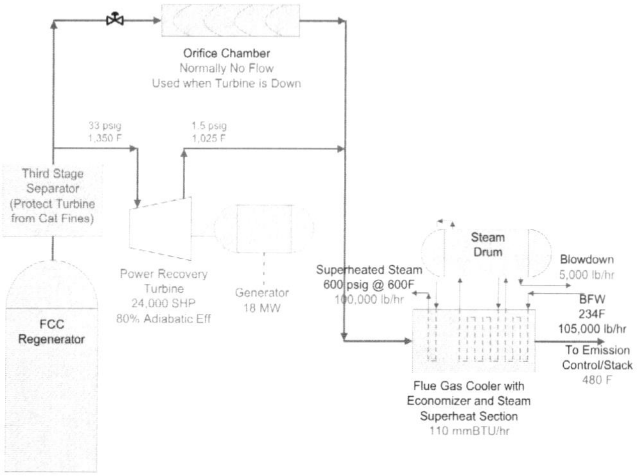

Figure 17: FCC flue gas power recovery revamp.

A revamp of the FCC regenerator includes a power recovery turbine and a third stage separator to protect the turbine from the catalyst fines. The hot flue gas is sent to the new power recovery turbine bypassing the orifice chamber and work is extracted. The power recovery turbine generates electrical power. The flue gas is cooled to 1,025°F at the outlet of the turbine and further cooled in existing Flue Gas cooler which will generate less 600 psig, 600°F HP steam.

For our case study example the electric power produced by the turbine offsets power generated at a coal fired plant. Depending on the type of projects identified to be eligible for offsets under the proposed Waxman-Markey bill, credits for the reduced GHG emissions could be provided to the refinery. A summary of all process data is provided in Table 4.

An installed cost estimate of $40 MM was developed for the revamp scope that includes the power recovery turbine, third stage separator and other associated installation items. It is assumed that sufficient plot space would be available for this revamp. The process schematic for the revamp is provided in Figure 17.

Limitations in the proposed Waxman-Markey bill that reduce the amount of the GHG credits received from offsets are accounted for in the economics. The IRR for GHG emissions prices from $0 to $100/metric ton for this project is summarized in Figure 18. The result shows that such a revamp would be marginal based on fuel savings alone. However with the GHG valued at $20/metric ton the threshold IRR of 20% would be achieved.

Table 4: Summary of FCC flue gas power recovery revamp results.

Generated		Base Case	Revamp Case
Steam	lb/hr	162,615	100,290
Electricity	bhp	0	24,340
Reductions			
Boiler Fired Duty	mmBtu/hr	0	-21.8
GHG Emissions Reduced			
Direct	metric ton/yr	0	-20,679
Indirect	metric ton/yr	0	161,387
Total	metric ton/yr	0	140,708

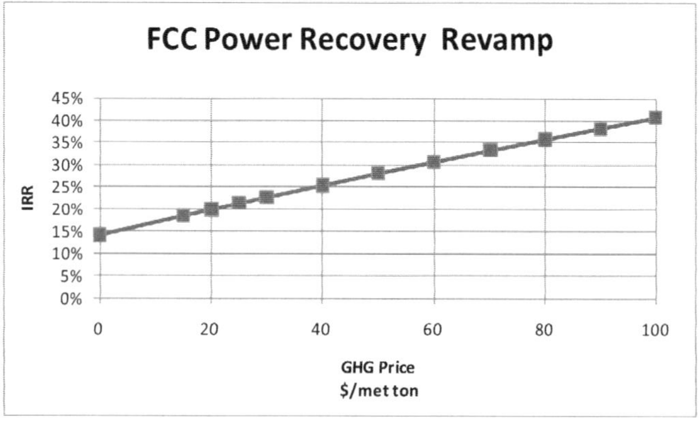

Figure 18: FCC flue gas power recovery revamp economics with GHG emission costs included.

3. Summary and Conclusions

The above case studies demonstrate a variety of refinery energy conservation projects that have significantly improved economics when the costs of GHG emissions from a cap and trade program are considered. As a result of installing all of the projects presented above a 9.5% reduction in GHG emissions from the notional refinery is achieved. A net summary of the reductions in energy use and GHG emissions are provided in Table 5.

Table 5: Summary of energy and emission reductions from proposed energy conservation projects.

Project	CO2 Reduced (1)	NG Reduced	Reduced Boiler Capacity	Power Utility Reduced	Offsets Required	Capital Cost
	metric ton/yr	SCFH	lb/hr	kWh		$MM
Crude Heater Inlet Temperature	34,161	91,758	-6,720	0	NO	19.5
Vacuum Tower Ejectors	6,649	16,376	10,679	-162	NO	2.8
Hot Unit Feed and Steam Generation	14,489	35,714	27,500	0	NO	3.2
FCC Power Recovery	140,708	-23,956	-62,325	18,158	YES	40.0
TOTAL	**196,007**	**119,892**	**-30,866**	**17,996**		**65.5**

(1) If offsets are required, CO_2 reduced includes the offsets awarded.

Observations from the various case studies show that care must be taken to evaluate all energy usage not only in the process unit of interest, but also changes in net refinery energy usage. Energy use in general will need a thorough review to ensure that all appropriate costs are included.

With the potential of offsets for power produced from coal, significant GHG reductions may be available from power recovery projects. Careful tracking of both electricity prices and available offset projects with the proposed legislation may provide some significant GHG reductions.

Various other opportunities are likely available at each refinery operation. An energy audit of the refinery operation is recommended so that the options presented as well as additional opportunities can be identified. The economic analysis should include the cost of GHG emissions since previously uneconomic projects may become viable as a result of proposed cap and trade for GHG emissions.

Acknowledgements

We would like to acknowledge the assistance provided by Graham Corporation for the analysis of the 3rd stage vaccum tower steam ejectors with a liquid ring compressor. We also acknowledge the budgetary costs for air preheat equipment provided by Tulsa Heaters, Inc. and OnQuest and typical economic factors for revamp and grassroot air preheat economics provided by Furnace Improvements.

References

1. U.S. Energy Information Administration (EIA), Emissions *of Greenhouse Gases in the United States 2008*, , U.S. Department of Energy, December 2009
2. National Highway Traffic Safety Adminastration (NHTSA), *Summary of Fuel Economy Performance*, U.S. Department of Transportation, March 30, 2009

(11b) Reduction of CO2 Emissions in a Refinery

Jason L. Stahlman, UOP, Des Plaines, IL

Richard Isherwood, Regional Services, UOP, Des Plaines, IL

The CO2 Emissions from a refinery may be reduced in a variety of ways. This presentation will focus on methods to optimize Utilities and Offsites related equipment; both inside and outside the process unit battery limits. By reducing fuel fired in Utility Boilers, and CO2 production in other areas of the refinery, the total CO2 emission footprint will be reduced. Specific topics that will be covered, include optimization of an MP steam header pressure, steam generation, multiple BFW headers, HRSG supplemental firing, and tank vapor recovery

See more of this Session: CO2 and GHG Issues Related to the Refining Industry

See more of this Group/Topical: Topical 7: 13th Topical on Refinery Processing

Update on Carbon Capture Technology

AIChE 2010 Spring Meeting

R.G.Hilton , Alstom

March 21-25, 2009

ALSTOM

Agenda

ALSTOM

- Overview
- Chilled Ammonia Process
- Advanced Amine Processes
- Oxy-Combustion Processes
- Summary

The Alstom Group:
a worldwide leader in its activities

ALSTOM

Total orders 2008/09: € 24.6 bn

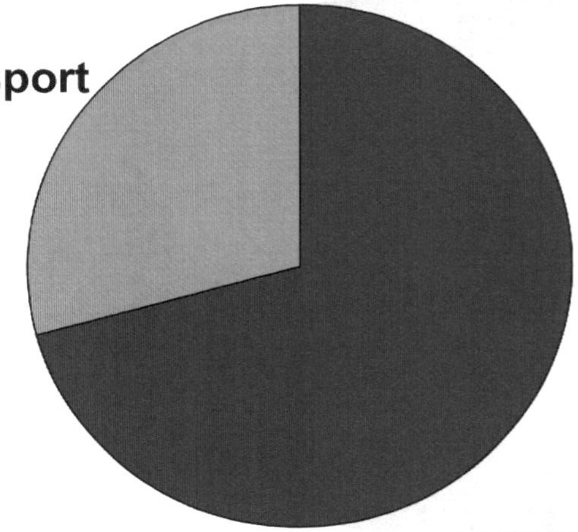

Coal, Gas, Hydro, Nuclear, Wind =
Full Power Systems Portfolio

N°1 in hydro power

N°1 in integrated power plants

N°1 in conventional nuclear power island

Recent acquisition of wind power

N°1 in air quality control systems

N°1 in services for electric utilities

Main Partnerships & Projects Portfolio

ALSTOM

Technology	Project	Size	Partners
Advanced Amines	Joint development programme West Virginia Pilot (USA) - Coal		Dow ALSTOM
	Belchatow (Poland) – Lignite	>250 MWe	PGE ALSTOM
	Archer Daniels Midland (USA) – Coal		ADM ALSTOM
Chilled Ammonia	Pleasant Prairie (US) – Coal	5 MWt	EPRI We ALSTOM
	Karlshamm (Sweden) – Oil/Gas	5 MWt	e·on ALSTOM
	Montainer (US) – Coal	58 MWt	AEP ALSTOM
	Mongstad (Norway) – Gas	40 MWt	TCM Catching our Future ALSTOM
	Mountaineer (US) – Coal	235 MWe	AEP ALSTOM
	TransAlta (Canada) – Coal	200 MWe	TransAlta ALSTOM
Oxy-combustion	Schwarze Pumpe (Germany) – Lignite	30 MWt	VATTENFALL ALSTOM
	Lacq (France) – Gas	30 MWt	TOTAL ALSTOM
	Alstom Boiler Simulation Facility Windsor (US) – Coals	15 MWt	NETL ALSTOM
	Jänschwalde (Germany) – Lignite	250 Mwe (Feasibility study)	VATTENFALL ALSTOM

6 Pilots in commissioning/operation

Tests completed

Agenda

ALSTOM

- Overview
- Chilled Ammonia Process
- Advanced Amine Processes
- Oxy-Combustion Processes
- Summary

Chilled Ammonia Process

ALSTOM

Principle

- Cooled flue gas is treated with ammonium carbonate in solution, which reacts with CO_2 to form ammonium bicarbonate
- Raising the temperatures reverses the above reactions – releasing pressurized CO_2

Advantages

- Energy-efficient capture of CO_2
- High CO_2 purity
- Tolerant to oxygen and flue gas impurities
- Stable reagent, no degradation possible, no emission of trace contaminants
- Low-cost, globally available reagent

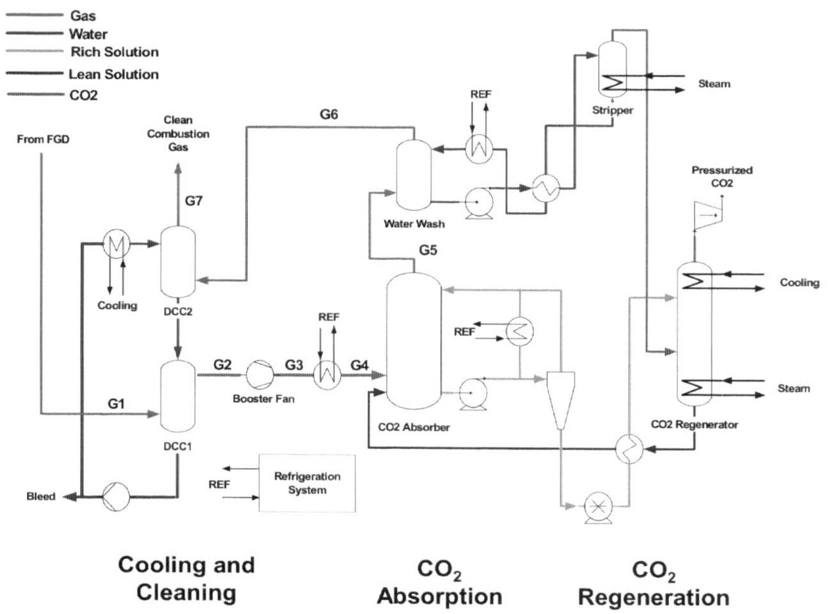

Chilled Ammonia Process
Summary Development Plan

ALSTOM

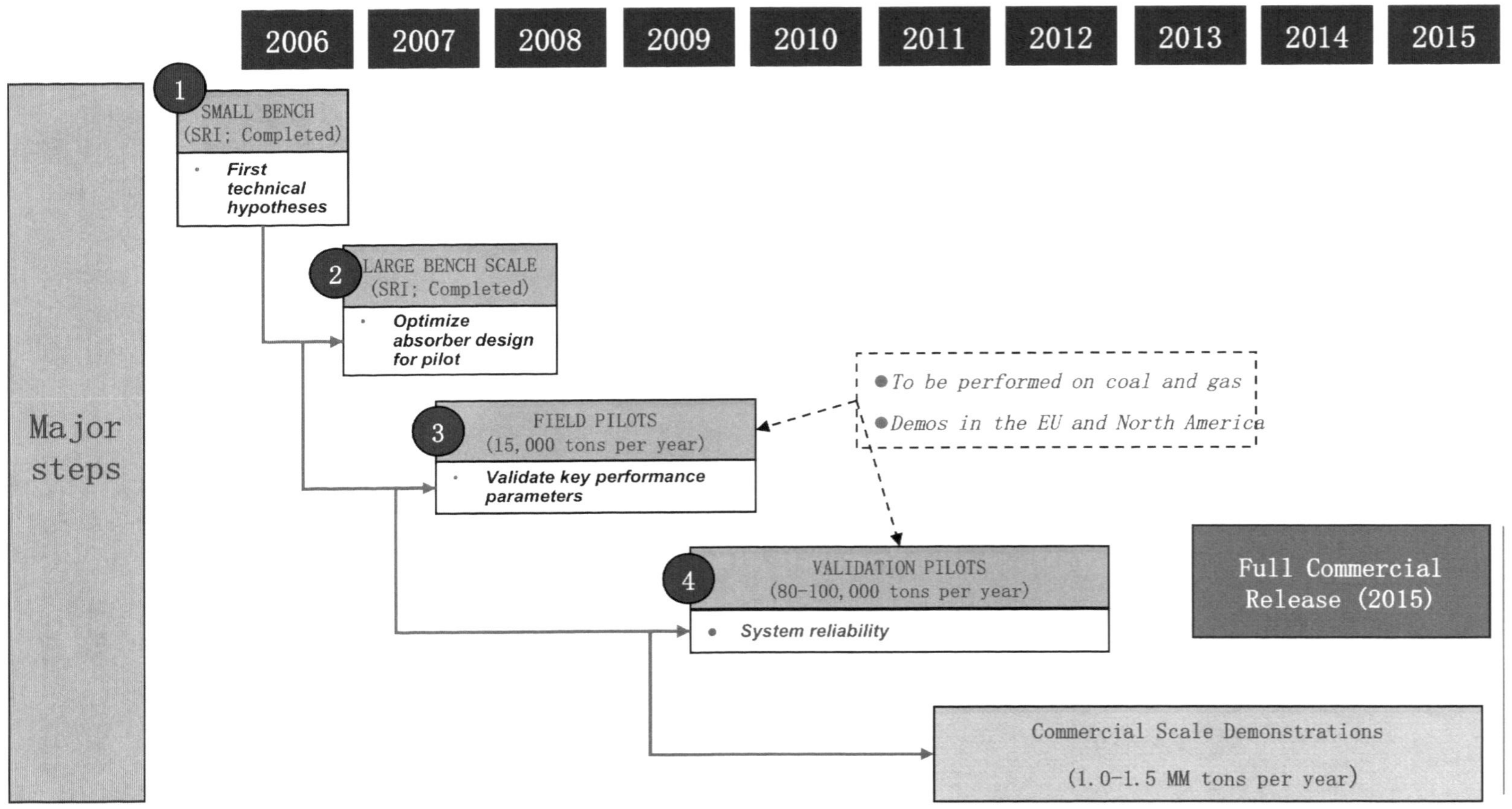

Basic concept feasibility to be validated as quickly as possible

Chilled Ammonia Process
Field Pilot at We Energies

ALSTOM

Pleasant Prairie Power Plant ("P4")

CO_2 Field Pilot at Pleasant Prairie

Validation Pilot Program

- Project participation through EPRI by 37 US and international utilities
- Designed to capture 1,600 kg CO_2/hour
- Operations commenced in June 2008
  - Over 6,500 hours operation
  - Achieved 90% capture in continuous operation at full load
  - Empirical data collected from pilot to date supports "proof of concept"
  - Initial data on steam and electrical energy consumption is consistent with

A great technical success!

Field Pilot Activities
Operating History

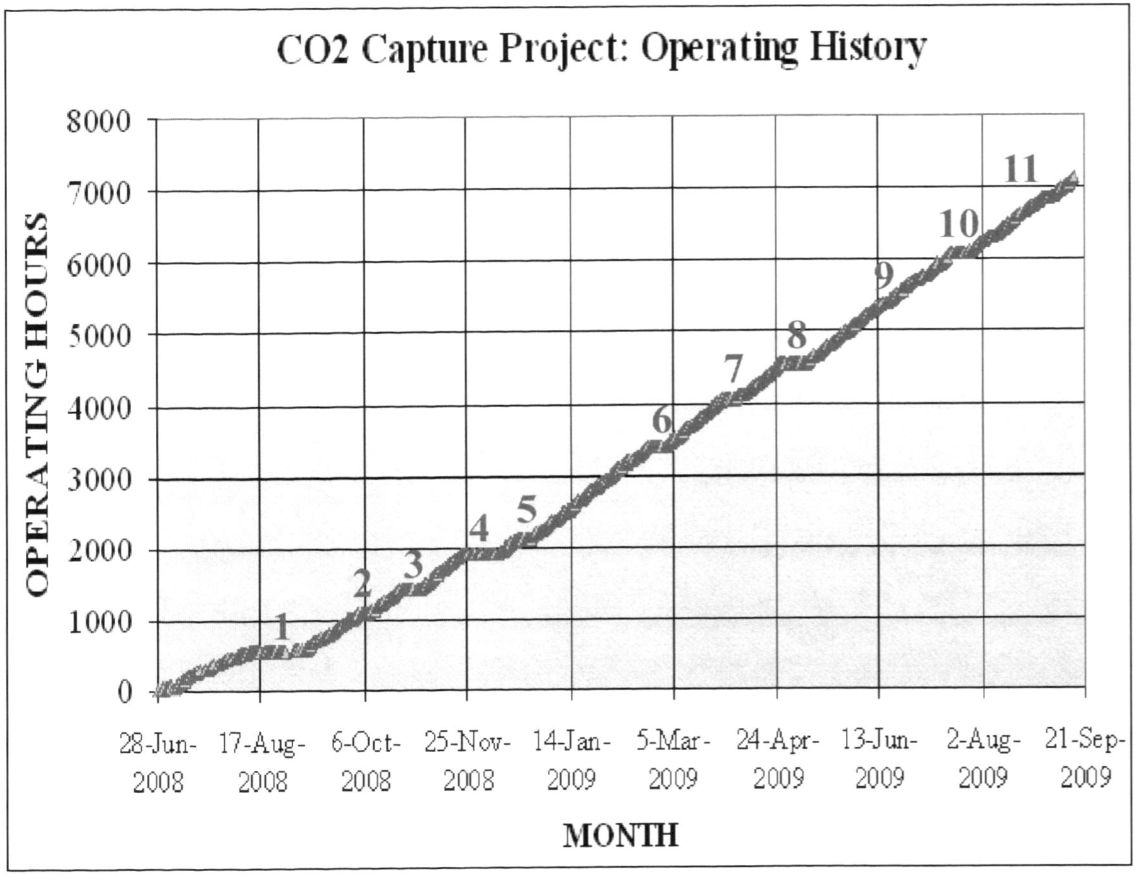

1. **Planned outage to support pilot modifications**
2. **Power Plant Forced Outage**
3. **Planned outage to support pilot modifications**
4. **Thanksgiving + Planned outage to support modification**
5. **Christmas Holidays**
6. **Power Plant Forced Outage**
7. **Planned outage for pilot maintenance**
8. **Unplanned outage for pilot maintenance (chiller)**
9. **Unplanned outage for pilot maintenance (stripper)**
10. **Planned outage to support piping modifications for Hydro cyclone and Absorbers**
11. **Planned outage to install additional instrumentation**

Total operating hours through Sept 16, 2009: 7,136

Validation Pilot at We Energies

CO_2 Quality

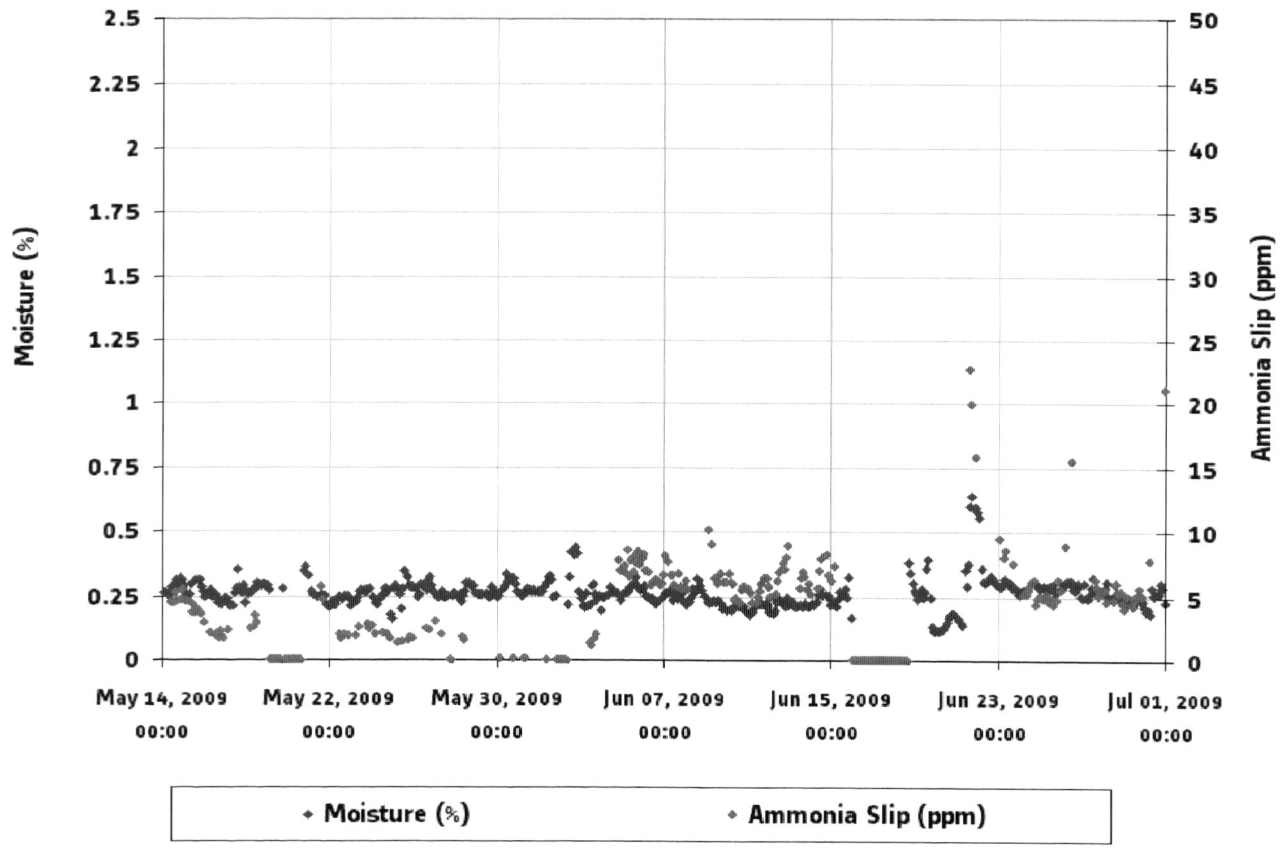

Validation Pilot at We Energies
3-D View of Pilot

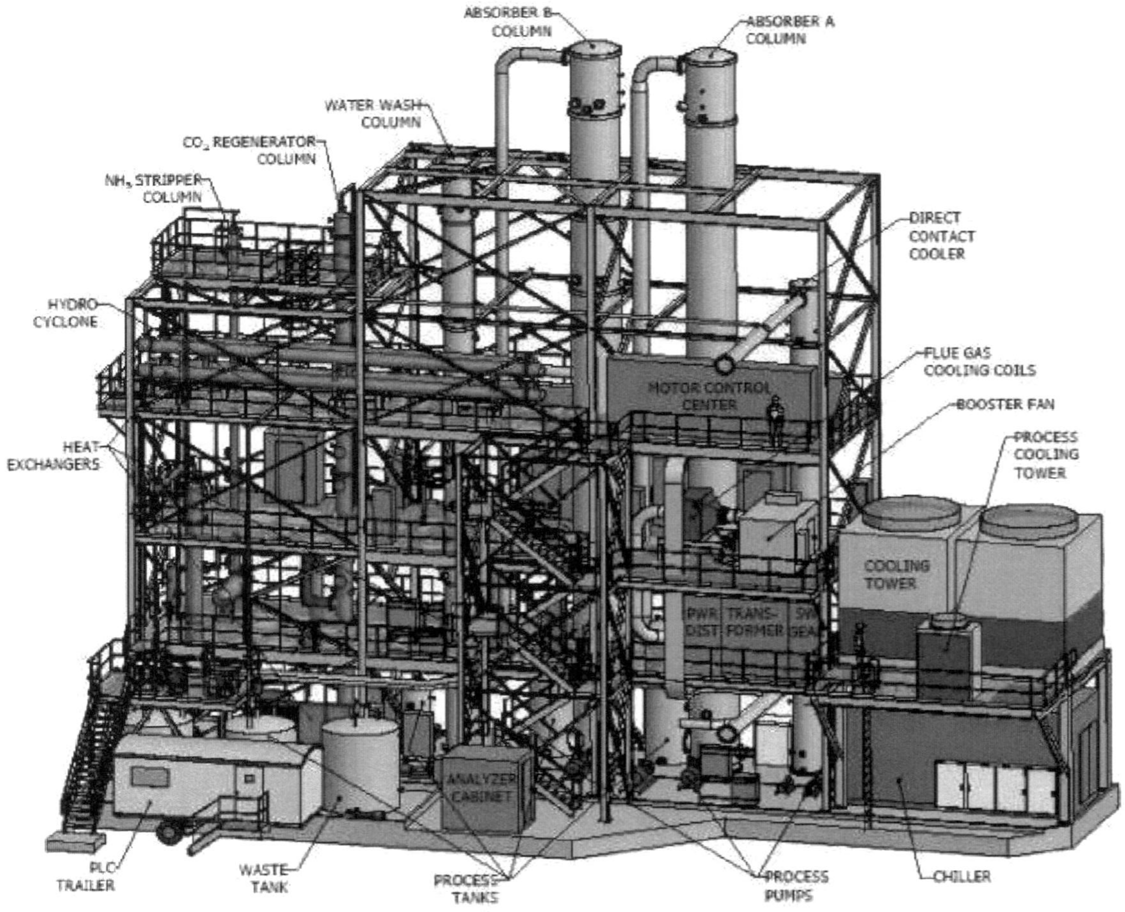

Validation Pilot at We Energies
CO_2 Capture Efficiency

ALSTOM

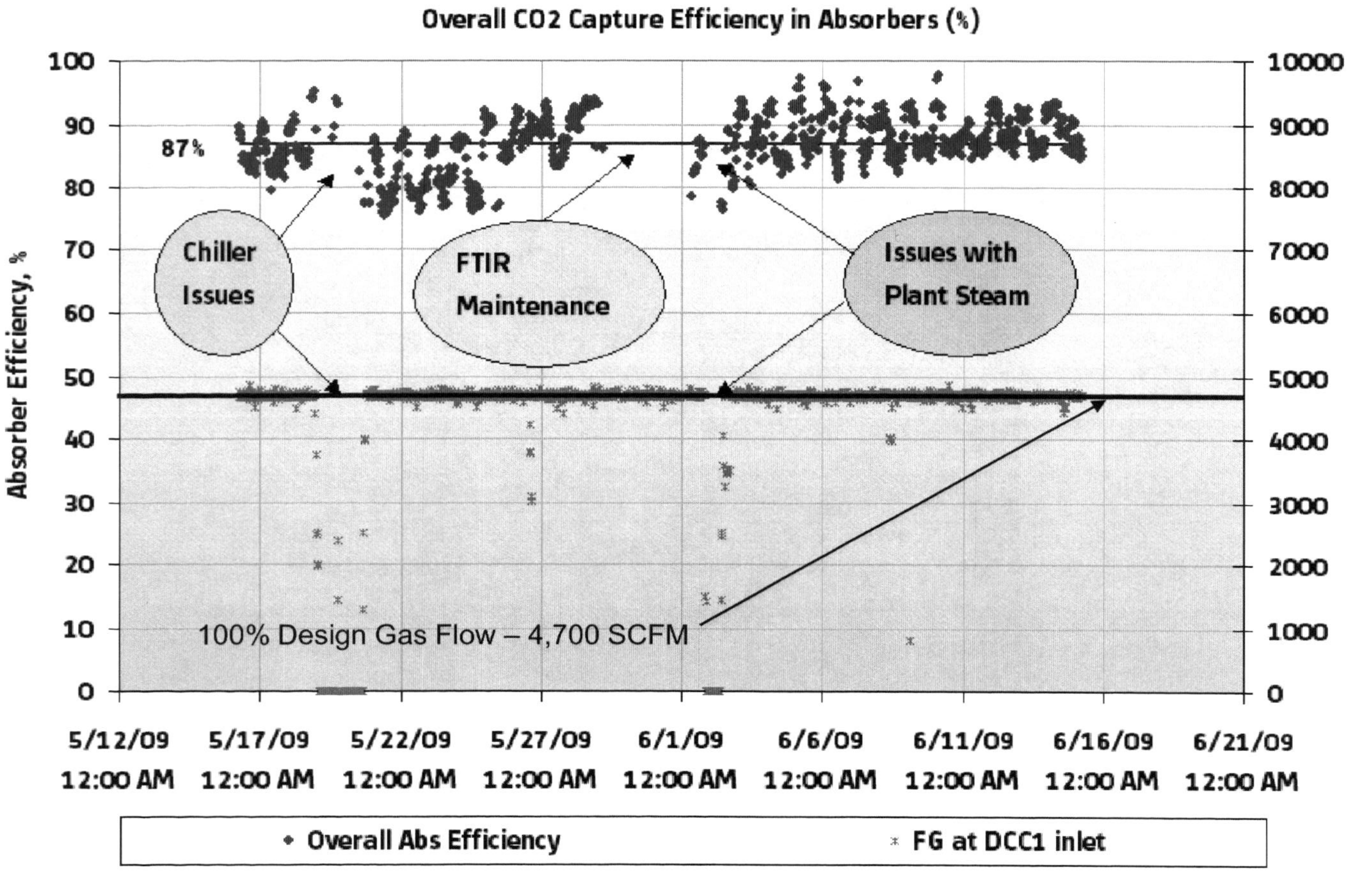

EPRI Material Monitoring Approach - 2009

ELECTRIC POWER RESEARCH INSTITUTE

- Every-day activities
  - Routinely collect and analyze liquid samples for NH_4^+, HCO_3^-, CO_3^{2-}.
  - Log FTIR gas analyzer data for CO_2, SO_2, NOx, NH_3.
  - Log other continuous data; T, P, F
- Multiple "mini-campaign" tests for material and energy balances:
  - Flue gas inlet — CO_2, O_2, N_2, NH_3, H_2O, Flow, T, P
  - Flue gas outlet — CO_2, O_2, N_2, NH_3, H_2O, Flow, T, P
  - CO_2 produced — CO_2, O_2, N_2, NH_3, H_2O, Flow, T, P
  - DCC blowdown — NH_3, Flow
  - Thermal / electrical / CW utilities
- One **campaign test** for material and energy balances around selected control volumes: (Manual) Gas, Liquids, and Solids sampling – Late Summer 2009.

EPRI Mini-Campaign Testing

ELECTRIC POWER RESEARCH INSTITUTE

Testing Objective – Accumulate high-reliability data on process performance.

- Independent Parameters:
  - Flue Gas Flow rate
  - Solution Strength
  - Process Temperatures, Pressures, Flows
- Dependent Parameters
  - CO_2 Removal, Quality
  - Utilities usage (Heat, electrical, cooling water)
  - Materials usage/disposal
    - Water CW blowdown / CW make-up
    - NH_3 loss / make-up

Mini-Campaign Sampling

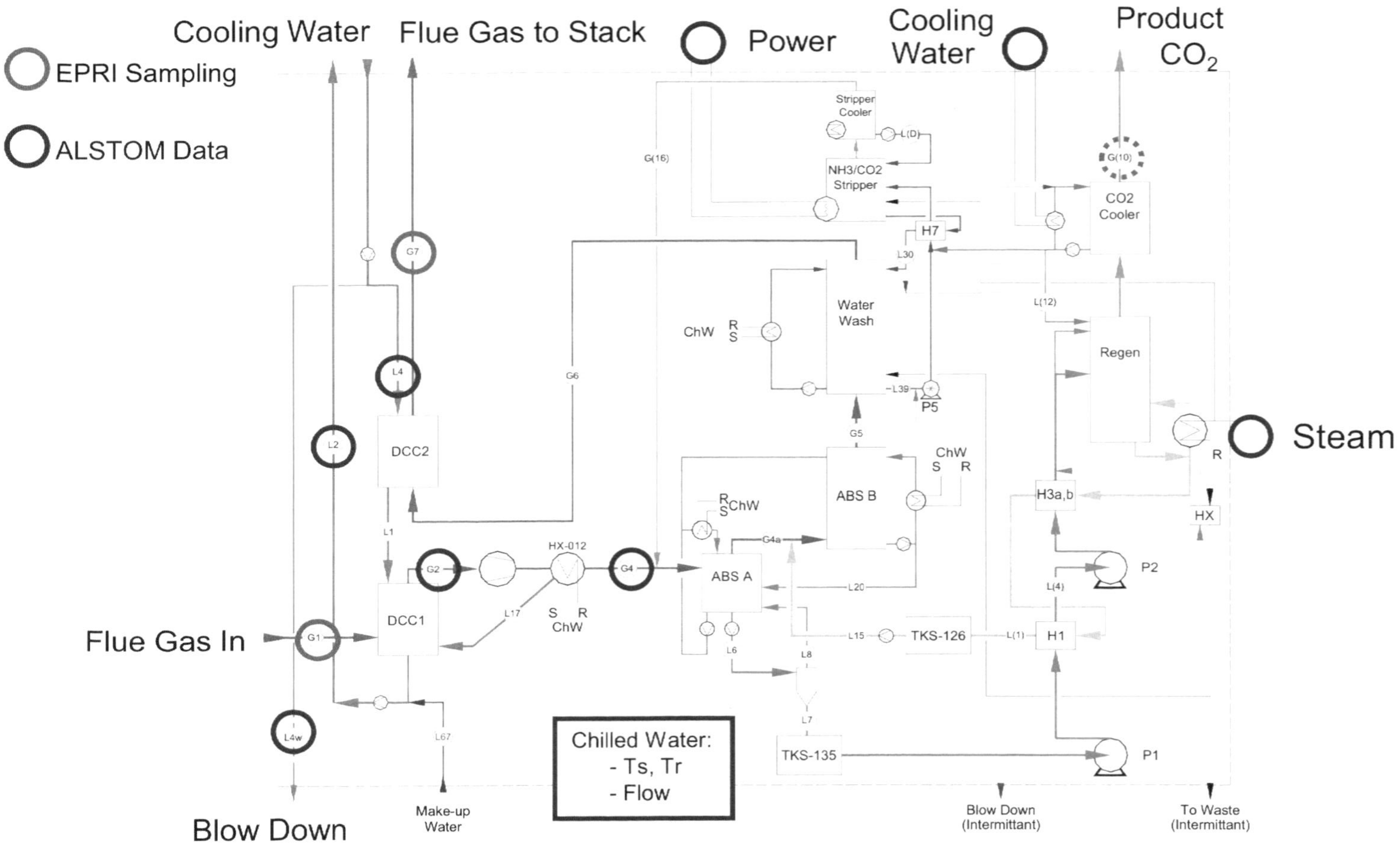

Mini-Campaign Test Conditions

ELECTRIC POWER RESEARCH INSTITUTE

Date	Inlet Flue Gas Flow	Inlet Flue Gas Temperature	CO_2 content vol %, wet	Solution Strength
April 22	85.1%	127°F 53°C	11.9%	Low
May 28	94.8%	128°F 53°C	12.3%	Medium
June 4	100.1%	130°F 54°C	13.0%	Medium
June 30	82.4%	130°F 54°C	11.5%	Low
July 1	106.9%	132°F 55°C	12.0%	Low

Overall CO_2 Removal Performance

ELECTRIC POWER RESEARCH INSTITUTE

Date	Inlet Flue Gas Flow	CO_2 Removal
April 22	85.1%	87.9%
May 28	94.8%	82.7%
June 4	100.1%	85.0%
June 30	82.4%	86.2%
July 1	106.9%	78.5%

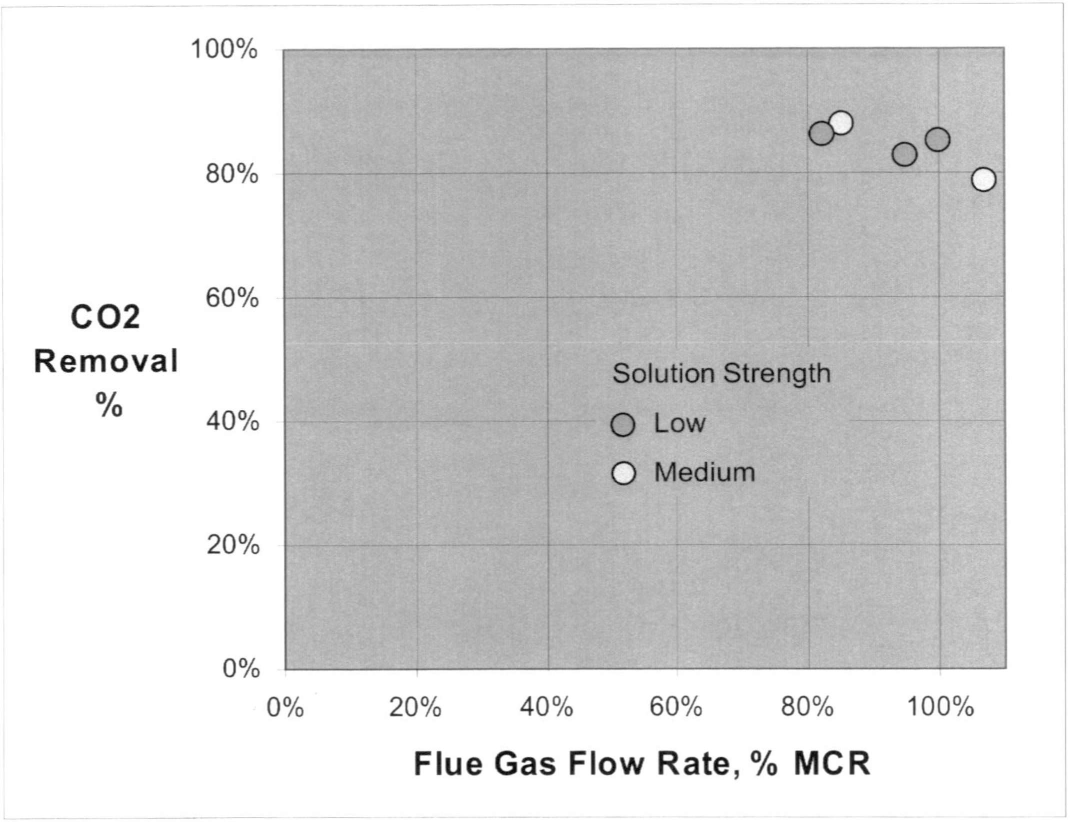

Ammonia Losses (Measured by EPRI)

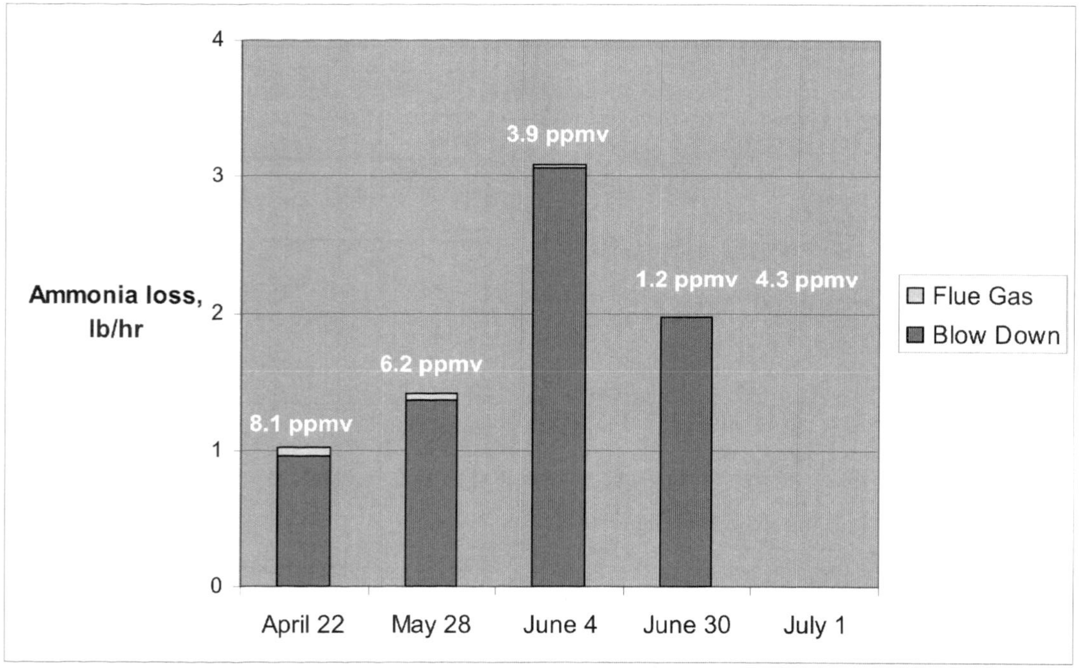

Ammonia loss streams

- DCC blow down
- Slip in flue gas
- Product CO_2
  much less than flue gas loss
- Cooling tower exhaust
  Insignificant when DCC cooling water pH is adequately controlled

Design blow down NH_3 loss is ~1 lb/hr.

July 1 blow down data still under review

Chilled Ammonia Process
Steam Energy Consumption

ALSTOM

- Steam Consumption
  - Heat of Regeneration
  - NH_3\CO_2 Stripping
- Heat of Regeneration includes:
  - Heat of reaction
    - Heat required to drive the reactions, regenerate reagent, produce pure CO_2 stream
  - Solids dissolution
  - Sensible heat requirements
  - Heat in overhead stream leaving the regenerator

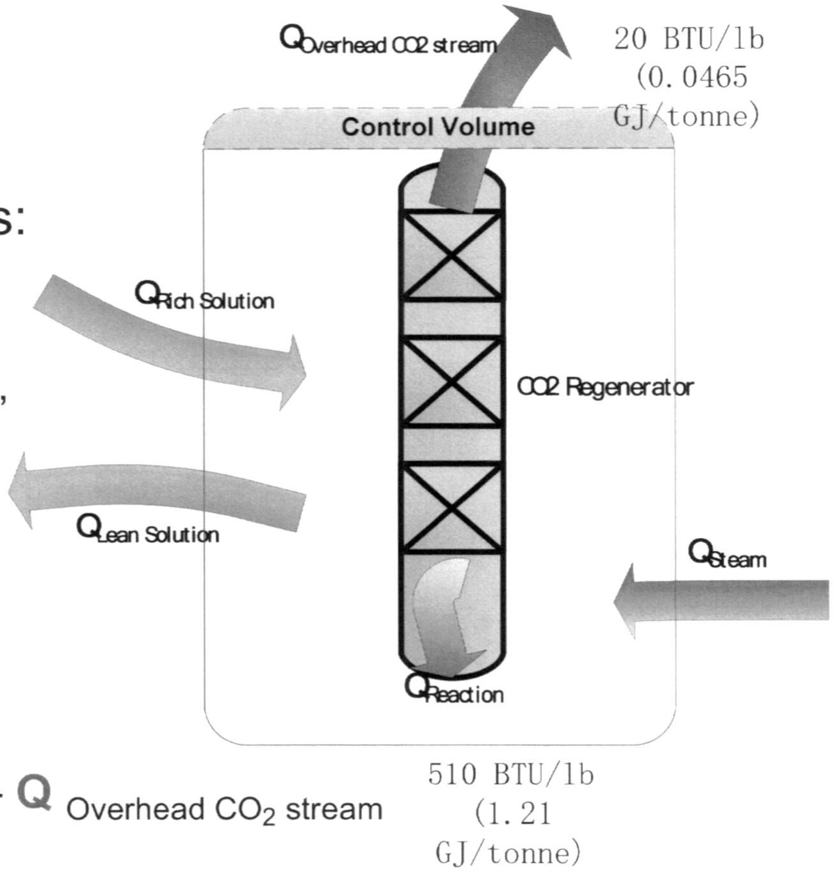

$Q_{\text{Heat of Reaction}} = Q_{\text{Heat Input (steam)}} - Q_{\text{Overhead } CO_2 \text{ stream}} - Q_{\text{Liquid Sensible Heat Losses (Lean/Rich solution)}}$

Validation Pilot at We Energies

Summary – Thermal Energy

ALSTOM

- Total steam energy consumption must be validated on an efficiently designed system that is demonstrated at a commercial scale and fully integrated with the power plant.
- In contrast, heat of reaction is dictated by the process chemistry and can be validated at the pilot scale.
- Heat of reaction measured at We Energies compares favorably with values determined in the laboratory setting;
  - Validates the figure used to size the validation pilots;
  - Validates the figure being used in commercial studies and FEEDs.
- The results extrapolated to a commercial scale system support Alstom's original estimate for total steam consumption

Validation Pilot at We Energies
Summary – Electrical Energy

ALSTOM

- Chiller system is primary component
  - Electrical demand varies with ambient conditions
  - Use of cold water source can eliminate chillers
  - Free cooling to minimize electrical consumption
  - Chiller load dependant on cross exchanger performance
- Estimated energy consumption for the pumps in line with expectations
- Booster fan draft loss lower than expected
- Total electric power consumption consistent with expectations, after adjustments to reflect a commercial design
  - Pump power demand based on typical efficiency for larger pumps
  - Refrigeration Coefficient of Performance (CoP) based upon levels provided by refrigeration suppliers for a commercial design versus the unit installed at We Energies

Accomplishments

ALSTOM

- Unit operations have been successfully integrated as a complete process
- Achieved high levels of CO_2 removal at design gas flow
- Accomplished high pressure regeneration
- Achieved low ammonia emissions
- Achieved high purity CO_2 with low concentrations of ammonia, water and other impurities.
- Steam and electrical energy consumption in line with expectations

Chilled Ammonia
Field Pilot at E.ON Karlshamn

ALSTOM

Karlshamn Power Plant (Sweden)

CAP Field Pilot

E.ON

Validation pilot program

- Designed to capture over 1,600 kg CO_2/hr
  - About 15,000 tonnes/yr at full capacity
- Unit commenced operations in April 2009
- Testing to continue for at least 12 months
- Located less than 60 miles from Alstom's Global Technology Center in Vaxjo, Sweden

Chilled Ammonia
Product Validation Facility at AEP Mountaineer

ALSTOM

AEP®

Technology Validation Program

- Designed to capture 100,000 tonnes CO_2/year
- Captured CO_2 will be sequestered into two wells on the plant property
- Commenced engineering and permitting in Oct 2007
- Started construction 2Q 2008
- Project start-up Sept 2009

Alstom's Chilled Ammonia Process at AEP's Mountaineer Plant, 5-21-09

All pictures of the Mountaineer CO_2 Capture and Storage Project are the property of Alstom Power and/or AEP

Mountainer Chilled Ammonia Plant

ALSTOM

Chilled Ammonia Process

Product Validation Facility at Technology Centre Mongstad

ALSTOM

Industrial demonstrator 80 000 tpy, Mongstad (Norway)

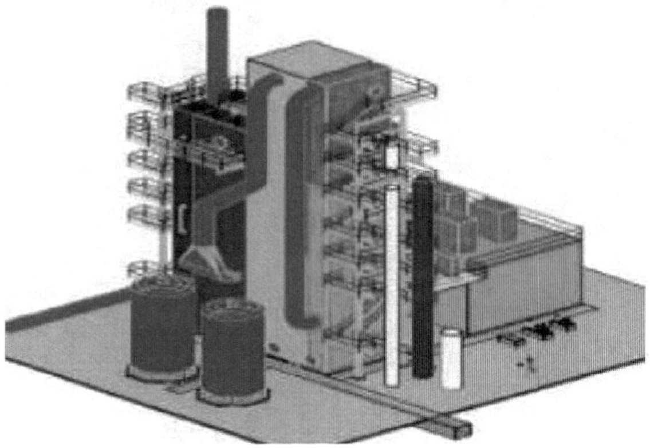

3D Model of Mongstad Pilot Plant

TCM Catching our Future

European CO_2 Technology Centre Mongstad

Technology Validation Program

- European CO2 Technology Centre Mongstad (TCM) for flue gases from natural gas CHP plant and a catalytic cracker (refinery)
- Designed to capture 80 000 t CO2/year
- Amine-based post-combustion to be tested in parallel
- Project schedule:
  - FEED study completed in November 2008
  - Detailed engineering on-going
  - Project approved by the Norwegian Parliament May 7th
  - Plant Commissioning complete: October 2011

Agenda

ALSTOM

- Overview
- Chilled Ammonia Process
- Advanced Amine Processes
- Oxy-Combustion Processes
- Summary

Advanced Amine Process
Technology Overview

ALSTOM

Principle

- Flue gas is contacted with an amine based solvent which reacts with the CO2
- Raising the temperatures reverses the above reaction – releasing CO2 and allowing the solvent to be recycled

Advantages

- Proven in natural gas & syngas purification
- Carbon capture from flue gas is a new application with new challenges
- More efficient capture of CO2
- Higher tolerance against oxygen & trace contaminants
- Lower solvent degradation rates

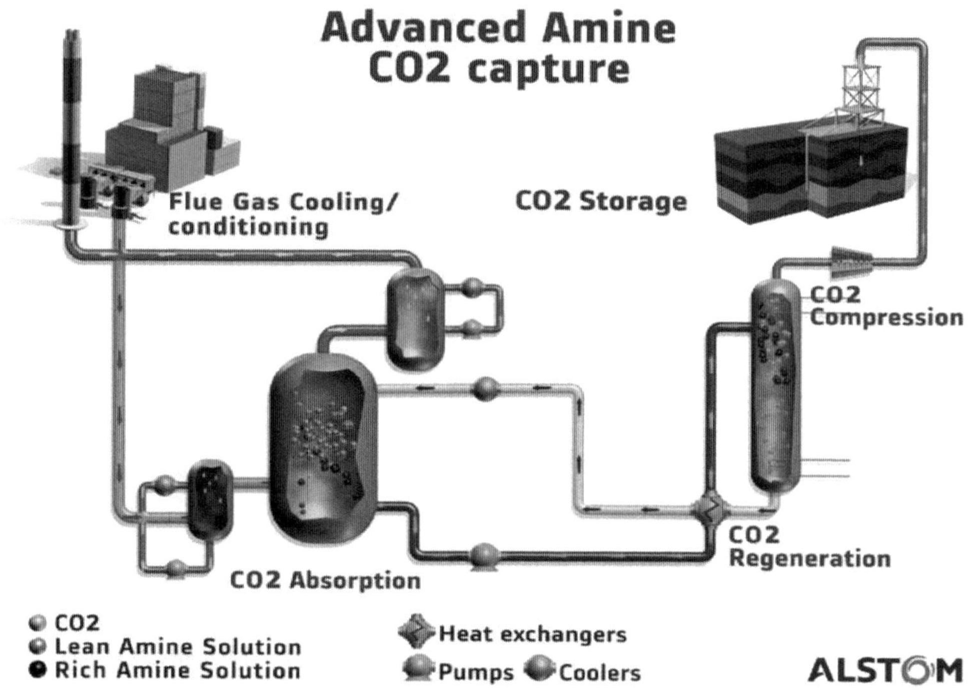

Source: Alstom

Joint Development
DOW ALSTOM

A strong team for a promising technology!

Advanced Amine Process
Field Pilot Testing at Dow Charleston, WV Facility

ALSTOM

Advanced Amine Field Pilot

Field Pilot Program

- Designed to capture 6-9 metric tonnes of CO_2/day
- Installed on both a coal fired boiler and a gas turbine at the Dow Chemical Facility in South Charleston, WV
- Project Schedule
  - Commissioning underway
  - Testing to start in September 2009
- Testing to continue into 2011

Archer Daniel Midland

ALSTOM

ADM

- Project Selected under US DOE ARRA Industrial Solicitation
- Phase I constitutes feasibility Study for CCS utilizing advanced amines
- Phase I commenced – projected for seven months to completion
- Partners: ADM, Dow, Schlumberger, and Illinois State Geological Survey
- Phase II: Based on outcome of Phase I, Capture and Sequester 2.2 million tons per year

Advanced Amine Process
Validation Facility - Belchatow

ALSTOM

PGE
Polska Grupa Energetyczna

Technology Validation Program

- MOU between Alstom and PGE Elektrownia Belchatow, signed Dec 08, Modified Jun 2009.
- CCS Facility proposed for EU funding
  - Capture greater than 1.8 million metric tonnes CO_2/year
  - Host facility 858 MW lignite unit under construction
  - Currently in pre-FEED for preparation of construction permit
  - Operational in 2014

Agenda

ALSTOM

- Overview
- Chilled Ammonia Process
- Advanced Amine Processes
- Oxy-Combustion Processes
- Summary

Oxy-Combustion Process

Principle

- Air Separation Units are used to produce pure oxygen stream
- Fuel is combusted in oxygen combined with recycled CO_2
- Resulting flue gas is concentrated CO_2 stream with impurities that are removed using conventional air quality control systems and gas processing units

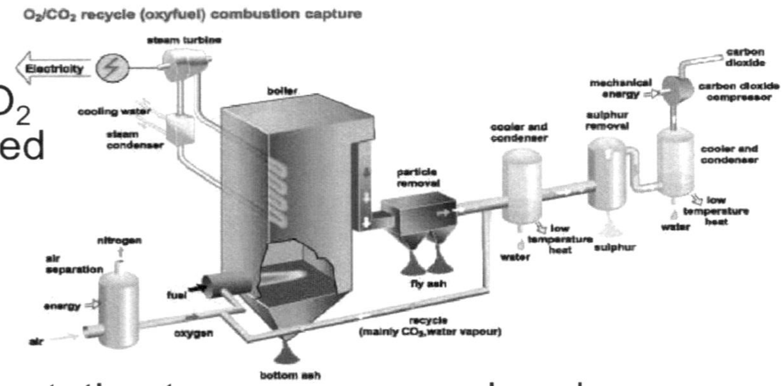

Advantages

- Reliability: components exist, only adaptation to power gen and scale-up
- All types of boilers / firing systems adaptable to oxy to cover complete fuel range
- Rapid scale-up to large size possible after demos. Retrofit in Oxy can be addressed
- Large panel of entities involved in development, contributing to reaching solid consensus

Current Status Oxy-PC
Schwarze Pumpe – 30 MWth Oxy Boiler island ALSTOM

VATTENFALL

(source: Vattenfall)

Goal

- Validation of oxy-fuel process, started Sept 08

Main features

- Boiler 30 MWth 40t/h
- Coal 5,2 t/h lignite
- Cycle 25b/330°C
- Oxygen 215 t/day
- CO_2 11 t/hr
- Alstom scope: Oxy-boiler + ESP

Boiler started Sept 08 - PAC Dec 08 – Testing on-going

Oxy-Combustion
Demonstration 30 MWth: TOTAL Lacq

ALSTOM

TOTAL

Demonstration/Validation program

Oxyfuel process starting 2008

- Retrofit of an existing boiler

Main features

- 40 t/h steam,
240 t/day Oxygen
- 150,000 tons CO2 will be
stored in a depleted gas field

30 MWt demonstration – Lacq, France

Source: TOTAL

First CCS with pipeline transportation & storage in Europe

Alstom activity on 12 major demonstrations

ALSTOM

Operating

Vattenfall Schwarze Pumpe
Germany - 30 MWth
Oxy - Lignite

AEP Mountaineer
USA - 58 MWth
Chilled Ammonia - Coal

EoN Karlshamn
Sweden - 5 MWth
Chilled Ammonia - Fuel

Total Lacq
France - 30 MWth
Oxy - Gas

Dow Chemical Co.
USA, West Virginia
Advanced Amines

Alstom BSF Windsor
US - 15 MWth
Oxy - Coals

Coming

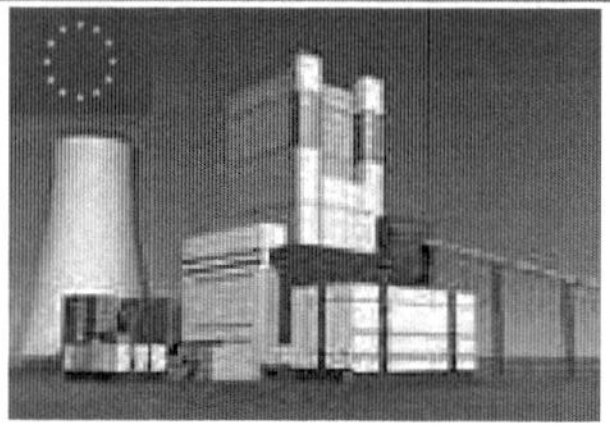

PGE Belchatow
Poland – 260 MWe
Adv. Amines - Lignite

Vattenfall Jänschwalde
Germany - 250 MWe
Oxy - Lignite

Statoil Mongstad
Norway - 40 MWth
Chilled Ammonia - Gas

Transalta
Canada - >200 MWe
Chilled Ammonia - Coal

ADM

Archer Daniels Midland
USA, Illinois
Advanced Amines - Coal

AEP Mountaineer
USA – 235MWe
Chilled Ammonia - Coal

Invited by the Commission to negotiate award agreements for EEPR funding

Selected by Alberta and Federal Canadian funding

Selected by US DOE to receive CCPI Round 3 funding

www.alstom.com

(11d) Aim for Autothermal Operation in Diesel Production

Francois R. Reverdy, Alfa Laval Packinox, Houston, TX

We are currently proposing to a Refiner that they modify a DHDS soon to be built, such that they will no longer need a fired heater other than for start up. We propose to modify slightly the heat integration and to improve the feed preheat temperature to the point that the system will work off the reactor exotherm. One of the selling points is that Heater emmissions will be just about eliminated .... The concept of autothermal hydrotreater operation is well proven for highly exothermic feedstocks like pyrolisis gasoline. New advances in heat transfer allow the same for diesel desulfurization.

See more of this Session: CO2 and GHG Issues Related to the Refining Industry

See more of this Group/Topical: Topical 7: 13th Topical on Refinery Processing

(11e) Screening CHP Configurations for Improved Performance and CO2 Reduction: Revisiting the R-Curves

Oscar Aguilar, Shell Global Solutions (US) Inc., Houston, TX

R-curves (i.e. performance vs. Power-to-Heat ratio plots) have been traditionally employed to characterize different designs of combined heat and power (CHP) plants. Conventionally such graphs are built assuming too simplistic arrangements that do not take into account the economics associated to each case.

In the present work, the R-curves have been revisited to handle complex configurations that can represent real CHP plants (e.g. in petrochemical sites) more accurately. In addition, economic curves (E-curves) and CO2-curves have been also developed to evaluate the cost and emissions implications of different designs interacting with external utilities sources (e.g. a power plant).

The main applications of this approach include screening CHP configurations for grassroots projects (design), targeting the performance, emissions and cost of existing plants (retrofit), and can be adapted to comparatively evaluate what if scenarios (operations). In all cases, time and effort can be reduced through a systematic (guided) approach to identify promising alternatives in design, retrofit and operational cases.

Keywords: Power Heat Ratio, Cogeneration, Energy Efficiency, CO2 reduction, targeting

See more of this Session: CO2 and GHG Issues Related to the Refining Industry
See more of this Group/Topical: Topical 7: 13th Topical on Refinery Processing

AIChE Paper Number 12a

CO_2 FOOTPRINTING IN THE CHEMICAL INDUSTRY

Russell Heinen
Vice President
SRI Consulting

Prepared for Presentation at the 2010 Spring National Meeting
San Antonio, TX, Date of Meeting (March 21 – 25, 2010)

CO_2 FOOTPRINTING IN THE CHEMICAL INDUSTRY

Russell Heinen
Vice President
SRI Consulting

Abstract: Retailers, customers and regulators – they all want to know about carbon footprints. And probably no single material ranks higher on their wish list than the world's largest-volume plastic: polyethylene.

As this paper shows, the footprint of producing polyethylene can vary considerably. So much so that in some circumstances a plastic bag shows a lower footprint than its paper analog, and in other circumstances not. And this is not just a deviation in physical factors such as feedstocks used; technologies applied, grades produced and geographic location, but also in methods used to calculate the footprint. Meanwhile, the near-term prospect of sugar-cane-based polyethylene from Brazil further complicates the footprint competition.

SUMMARY

Produced at 70 million tonnes per year, polyethylene is all over the place. So, too, is its carbon footprint, which under certain conditions even can be negative.

In this age of 'inconvenient truth', massive popularity and attention have accrued to carbon footprinting. At the heart of footprinting is a competition, a rivalry to decide which products – automobiles, buildings, foods, fuels, packages or whatever – are lowest-carbon. In many sectors, particularly building and packaging, polyethylene (PE) is a significant part of the footprint. The stuff is everywhere, well-known and surely the chemical industry's most-recognized product.

Well-known, unfortunately also well-loathed. Laws to discourage or ban the use of plastic bags have been enacted in countries including Australia, Bangladesh, Ireland, Italy, South Africa and Taiwan, and cities including Los Angeles, Mumbai and San Francisco. This has gone on for years, and the story gets worse. Only this June, the Executive Director of the UN's Environment Program, Achim Steiner, reportedly called for 'single-use' plastic bags to be banned globally because of their degrading effects on the planet.

Given such criticism, now is an opportune time to examine PE's carbon footprint, which according to public opinion is one of the most degrading effects on the planet. So we recently crunched the numbers, calculating the carbon shoe-size of the three main types (HD-, LD- and LLDPE) from leading ethylene feedstocks (naphtha, ethane and so on) in four regions (China, Europe, Middle East and the US).

So what is polyethylene's footprint? As the paper shows, it lies somewhere between 1 and 3 tonnes of CO_2e per tonne. This is an enormous variation, enough to change footprint-based decisions. Behind this are three commercial reasons, plus one – even larger – accounting reason.

The three commercial reasons for footprint variation each create about a 1.5:1 span from highest to lowest footprint:

- *Ethylene feedstock* – a rough ranking, from highest to lowest footprint of the main feedstocks is: gasoil, naphtha, ethane and ethane/propane.
- *Region of production* – China shows the highest footprint, the Middle East the lowest with the US and Europe in-between.
- *Type of polyethylene and polymerisation process* – low-density and linear-low-density grades show a higher footprint than high-density grades, with the LLDPE slurry process coming in highest.

Yet the largest variance in footprint is caused by the allocation method, the rules by which the footprint of the steam cracker (used to make ethylene) is apportioned among its products. Changing from one allocation method to another can vary the footprint of polyethylene by as much as 2:1

The first three reasons for variation are obvious grounds for competition. However allocation is a continuing source of confusion. There are four common allocation methods in use (see figure) that deliver radically different answers. So our paper's first recommendation is that the industry tries to adopt a single standard for allocation, and we suggest it be the mass High-Value-Chemicals (HVC) method that has developed and used by the International Energy Agency in its studies of ethylene and other petrochemicals.

The mass HVC method combines the convenience and reproducibility of both economic- and mass-based methods (recommended respectively by BSI and ISO), and it most accurately reflects decision-making in the real world (e.g. that companies operate steam crackers to add value, not to produce fuels).

Figure 1
ALLOCATION METHODS CAN CHANGE PE'S FOOTPRINT DRAMATICALLY
(MIDDLE EAST, BY FEEDSTOCK)

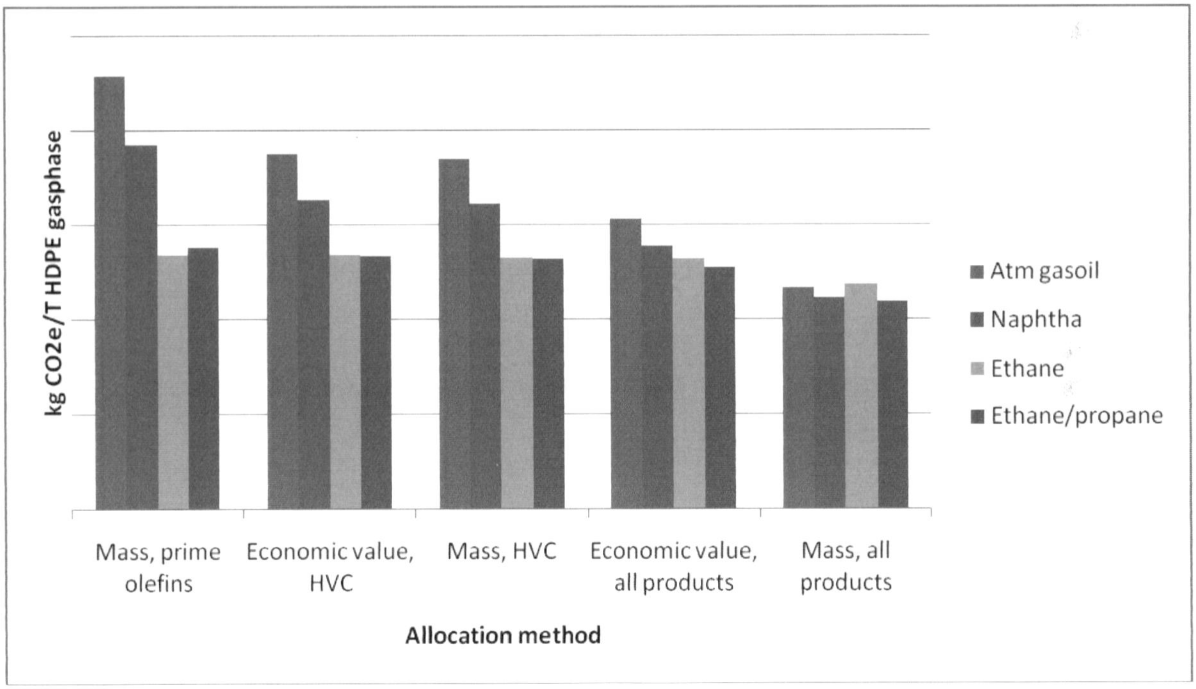

How well does polyethylene compete? Not surprisingly, this depends on which footprint variant is chosen. At the lower end of its range, when faced with the ubiquitous competition versus paper and a biodegradable-plastic in carrier bags, polyethylene wins. At the higher end of its range, it loses to the biodegradable and is roughly equal to paper. So the second recommendation of this paper is that economic players in the polyethylene chain find out where they rank.

INTRODUCTION

Fuelled by society's ever-increasing concern about climate change, carbon footprinting has become very popular.

Retailers and suppliers are under increasing pressure to report the carbon footprints of their products to their customers and regulators. Such suppliers include makers or sellers of: automobiles, buildings, electronics, food, fuels, heating systems, office supplies, packaging, personal-care and household-care items and probably others. In response, retailers, suppliers and regulators are asking supply-chains for carbon-footprint data. An implicit threat and opportunity is that they will shun products with high footprints and favor products with lower ones.

This paper presents a production footprint for polyethylene, deconstructs it, shows its variations and compares it to other estimates. It then shows how this variation can affect downstream decisions, using a comparison of polyethylene and paper shopping bags. The primary 'actionable information' is that anybody in the supply chain, faced with requests for footprint data, can use this Footprint Report to decide how to respond.

As explained, this is not an academic exercise. Footprints are being used to make decisions. Chemical companies are being asked for footprints of their products. Polyethylene is a good product to analyze, because it is one of the industry's largest volume products, and because it is a bellwether of the industry's environmental performance. This latter point is demonstrated by what over the past decade has become a common occurrence: a proposed or actual ban on plastic bags.

Footprints in Action: Guiding Public, Commercial and Personal Decisions

Carbon footprints are being used to make decisions. The most significant application is in setting fuel taxes. In 2007 the developed world (i.e. the OECD countries) paid $13-15 thousand million (US billion) in subsidies and tax relief for biofuels primarily in the belief that these have lower carbon footprints than conventional fuels.

In the United States, the cost to taxpayers of just the federal volumetric tax credits for biofuels is estimated to be almost $ 4 billion in 2007, equivalent to one third of the total $12 billion expected to be paid out in farm support in 2007. Federal tax credits for biofuels could grow to $16 billion if the US Congress were to adopt proposed expanded 'alternative fuels' target of 35 billion gallons (132 billion liters) a year by 2017. In the European Union, reduced excise tax rates for biofuels are estimated to have cost around Euro 3 billion in tax revenues foregone in 2006, up from Euro 1.8 billion in 2005.

The EU and its member states are also using footprints to set vehicle taxes, to define building codes (for 'low-carbon' construction) and to shape national policies on energy, transport and housing.

Major retailers such as Office Depot, Tesco and Walmart are starting to offer so-called *carbon labels* on some of their merchandise. The French government plans to introduce mandatory carbon labels on a broad range of products in 2011. Other countries are watching closely and planning similar schemes. Organizations as diverse as the Barclays Bank, International Football Federation (FIFA), KPMG, Marks & Spencer, Nike, Timberland and Yahoo! claim to have become or that they will become carbon neutral.

One quantitative indicator of this commercial/personal trend is the burgeoning market for voluntary carbon offsets. To counterbalance their greenhouse gas emissions, Europeans and North Americans in 2007 spent as much as $200 million on voluntary offsets and by 2010 they may spend up to $4 thousand million (World Bank 2007).

Footprints of Chemicals Are In Demand

Chemical companies are in the supply chains (and decision-chains) that are being footprinted. Requests for footprints are coming in from customers and customers' customers. In response, chemical makers have two basic options, i.e. to report:

- *Their in-house data.* If they do this without any reference to the range of possible footprints in the market, they run the risk of looking bad. They also miss any opportunity of looking good.
- *'Consensus' figures.* Ok, but which ones, and what are the risks and opportunities? Why is there such a variance in possible footprints? Are there any consensus footprints? If not, which ones are right?

Meanwhile, companies frequently are asked to join in expensive data-collection exercises. What are the pros and cons? Is this really necessary?

So, chemical companies need footprint benchmarks and explanations of the benchmarks, i.e. what could the footprint be, what are the range of footprints and why. Benchmarks and analysis could be generated as individual studies, but an off-the-shelf, multi-client report is far cheaper. For credibility's sake the work should be done by a third-party, and ideally, it will help to build consensus as to what footprints should be.

Why Polyethylene?

The greatest demand for footprints will be in high-volume, high-profile products that have high footprint variation, which makes polyolefins an obvious choice for paper.

Polyethylene is clearly high volume. With global, annual production topping 60 million tonnes, it is one of the industry's biggest products. It is the backbone of the ethylene chain (Figure 3), consuming some 60% of global ethylene, which at over 100 million tonnes per year is the industry's largest product. Of course this means that polyethylene is enormously important for the economic health of the chemical industry.

Polyethylene is very high profile: it is a symbol, an icon of the chemical industry. Partly this is due to its economic clout, but also this comes from its omnipresence – where can you go and not see polyethylene? Unfortunately, this is often seen unfavorably. Laws to discourage or ban the use of plastic bags have been enacted in countries including Australia, Bangladesh,

Ireland, Italy, South Africa and Taiwan, and cities including Los Angeles, Mumbai and San Francisco.

Finally, as the rest of this paper shows, the footprint of polyethylene can vary considerably.

METHOD OF THIS PAPER

Carbon footprints are a summation of the greenhouse-gas emissions of a product or service. A footprint is a measure of that product's or service's contribution to global warming, often referred to as climate change.

Footprinting methods applied in this paper are informed by footprinting standards and by previous studies. (For instance, we have applied standard GWPs as specified by the IPCC. The footprints presented are believed to represent the vast majority of polyethylene production, and to be far more precise than the only other known paper of its type (from Plastics Europe). Allocation method is critical to the result.

CARBON EMISSIONS IN (CONVENTIONAL) POLYETHYLENE PRODUCTION

Polyethylene is produced by polymerizing ethylene, sometimes with minor additions of other raw materials. The cradle-to-gate supply of polyethylene is described in the following segments:

- raw materials production
- ethylene-feedstocks production
- ethylene production – conventionally this is done by steam-cracking feedstocks derived from oil and gas (covered in this chapter); an alternative method (presented in *Carbon emissions for bioethylene from sugar cane*) is to dehydrogenate bioethanol made from sugar cane
- other raw materials production
- electricity generation
- polyethylene polymerization

The options for these (Figure 2) and their associated carbon emissions are presented in the following subsections.

Figure 2
POLYETHYLENE PRODUCTION, CRADLE-TO-GATE FLOWCHART

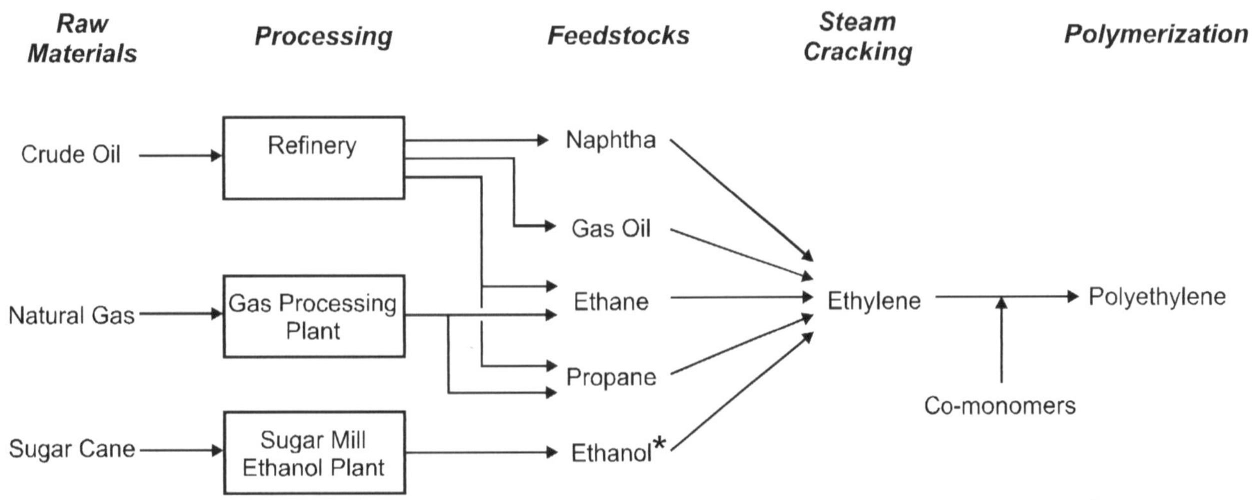

* Via Ethanol Dehdrogenation (Not Steam Cracking)

Conventional Steam Cracker Yields and Allocations

For application in this paper, SRIC's cracker yields – as traditionally presented in PEP reports – have been used.

In cost or economic analyses, the conventional presentation of a cracker's yield includes: 1) all inputs and outputs normalized according to ethylene production; 2) all products in mass units, except for methane-rich gas, which is presented in energy units (i.e. million calories, written as MMCAL); and 3) methane-rich gas as a gross quantity (i.e. it is not presented net of fuel consumed to fire the cracker itself

Steam-Cracker Greenhouse Gas Emissions, Unallocated

To define the greenhouse gas emissions of steam-cracking each of the feedstocks, we used figures from SRIC's Greenhouse Gases Handbook.

GHG emissions are of three types: process, fuel and electricity.

- *Process* emissions comprise all carbon containing waste streams leaving the process. These can be either vapor or liquid. Typical vapor streams would include vent streams, vapor purges, flared streams, and incinerator flue gases. Typical liquid streams would include aqueous streams that contain carbon-bearing components.
- *Fuel* combustion emissions fall into two major categories: those from furnaces and direct fired heaters, and those from steam generation.

- *Electricity* generation emissions are based on the US average of years 2001-2003 of 584 g CO_2e/kWh (0.584 tonnes CO_2e /MWh). Electricity is presumed to be generated 'over the fence', i.e. no allowance has been made for transmission losses.

For all feedstocks, fuel combustion dominates the GHG emissions profile, accounting for about 95% of the total. Process and electricity generation roughly split the remaining 5% of emissions.

This analysis – for ethylene – leaves out the impact of shifting from one electric power grid to another; the ethylene footprints are based on US power generation. However, in this case such variations are not important, because carbon intensity of power is not significant for ethylene. Shifting from the US to any of the main countries in the other regions changes the overall steam-cracker footprint by less than 1%.

Electricity Generation

For this paper, we have used footprints defined by EcoInvent (EcoInvent), which include not only the power plant operations but also the production, transport and waste-disposal associated with the fuel (i.e. a cradle-to-gate footprint for power generation). The four regions modeled in our analysis are: China, Europe, Middle East (Saudi Arabia) and the US.

Carbon emissions for electricity generation can vary dramatically by region; this is largely a function of the fuel mix used. For instance, electricity in Norway, Sweden and Switzerland, which rely heavily on hydropower, has much lower footprints than electricity in China, Germany or the US, where coal has a significant fraction of the fuel mix (Table 1).

Table 1
ELECTRICITY FOOTPRINTS, SELECTED REGIONS, AS REPORTED BY ECOINVENT

	Footprint	
Region	g CO_2e per MJ	g CO_2e per kWh
China	317	1140
Europe (UCTE)	136	489
Finland	122	439
Germany	184	662
Norway	2	8.1
Saudi Arabia	161	579
Sweden	11	38.9
Switzerland	5	19.1
UK	166	596
USA	209	754

A cogeneration footprint was not modeled as such. The main reason for leaving co-generation out: it is a minor factor in footprint differences – EcoInvent assigns cogenerated electricity 80% of the footprint of conventionally generated power. This is far less than the differences encountered by country. By contrast, cogenerated steam has a significantly lower footprint than conventional steam, but this does not play a meaningful role in the production of polyethylene.

Polymerization

Polyethylene is produced by polymerizing ethylene, sometimes with the addition of co-monomers. The markets define three main types of polyethylene: high-density, or HDPE; low-density, or LDPE; and linear-low-density, or LLDPE. These account, respectively, for roughly one-half, one-quarter and one-third of polyethylene output.

HDPE's footprint is lower than LDPE's and LLDPE's, while LDPE's footprint is in the lower-middle of the range of LLDPE's. Their footprint components – process, fuel and electricity – can vary substantially. Compared to the other grades, producing LLDPE can have relatively high process and fuel footprints, while polymerizing LDPE consumes significantly more electricity.

Data sources used for carbon footprinting of polymerization are internal to SRI Consulting: the GHG Handbook; Process Economics; and World Petrochemicals. Footprints, which have been calculated by process type, vary from as low as 178 kg CO_2e per tonne of polyethylene to as high as 761.

ALLOCATION OF STEAM-CRACKER FOOTPRINTS

In polyethylene production, the most difficult process to allocate is the steam cracker. Allocations for other processes are either unavailable (production of feedstocks), relatively insignificant (production of co-monomers) or easy (polymerization, where polyethylene is the only product, so no allocation is needed).

How should the GHG emissions from a steam cracker be allocated among its products? Below we review allocation methods suggested by carbon-footprinting guidelines, some principles suggested by common practice and precedent, and then the allocation results themselves.

Primary Allocation Methods

Based on guidance and practice, there appear to be four primary methods for allocating a steam cracker footprint among its products (Table 2) that result in three allocation profiles.

Table 2
POSSIBLE ALLOCATION METHODS
(alphabetical order)

Allocation method	Used or recommended by	Comment
Economic value, to all products	PAS 2050	
Mass, to all products	Plastics Europe	Recommended by ISO as an alternate method
Mass, to high-value chemicals	IEA	Recommended by ISO as an alternate method, preferentially to allocating to all products
Mass, to prime olefins (ethylene and propylene)	Chemical industry	

For prime olefins and HVCs, it is presumed that allocation is typically done on a mass basis. For allocation to all products, it is done either on an economic basis (recommended by PAS 2050) or on a mass basis (an alternative recommendation of ISO's). This is not to say that an economic basis cannot be used for prime olefins or HVCs, but that in practice it appears not to be.

In the figure below we present the results of using different allocation methods for three feedstocks used for producing ethylene. As shown the heavier feedstocks that have larger co-product yields show significant variations between allocation methods.

Figure 3
ALLOCATION METHOD IMPACT ON ETHYLENE'S FOOTPRONT

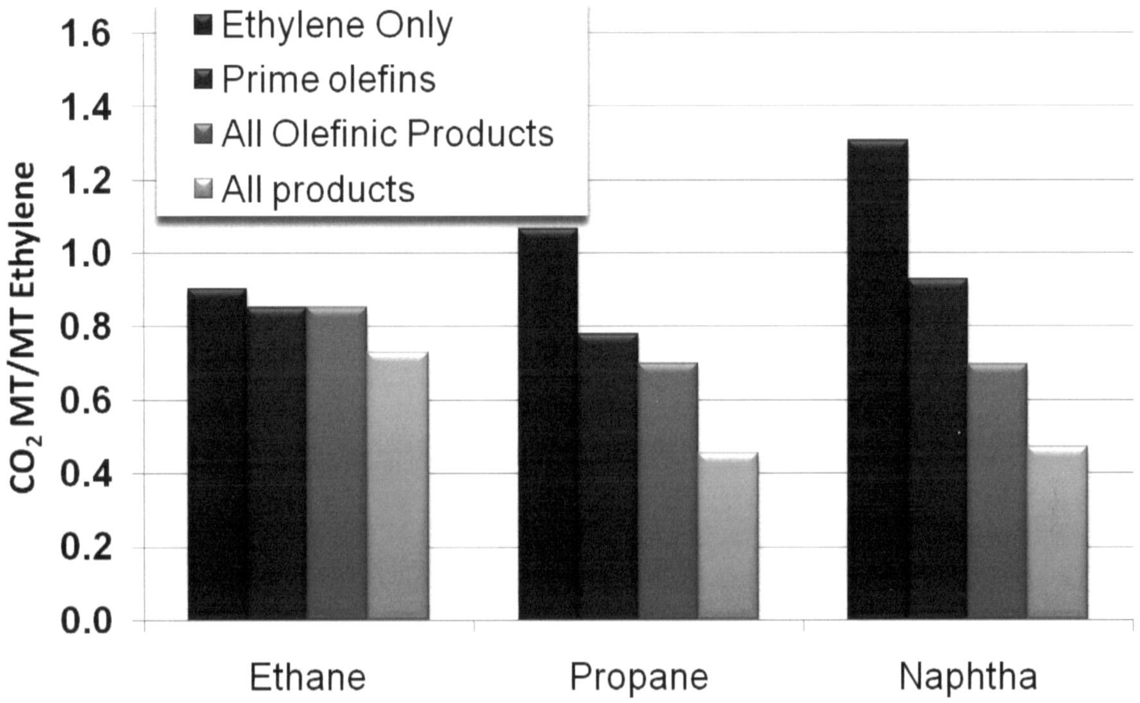

POLYETHYLENE'S CARBON FOOTPRINT – HOW IT WAS COMPILED

To show how the data from the preceding chapters are compiled into a footprint, in this case we present the calculation of an the footprint of HDPE (slurry process) – which accounts for about one-quarter of polyethylene production – from naphtha in Europe. The steam-cracker footprint allocation method applied is 'mass all products'.

The calculation is presented in two ways.

First, the footprint is shown by tonne of product (in Figure 4), the columns on the far-left and the second from the far-right). Production of crude oil delivered to Europe is 275 kg CO_2e/tonne. This is multiplied by 1.07, the allocation ratio of crude oil:naphtha, which yields 294 kg CO_2e, then added to the 158 kg CO_2e/tonne for naphtha refining, giving a total naphtha footprint of 294 + 158 = 452 kg CO_2e/tonne. This 452 kg CO_2e/tonne is multiplied by 1.0 and then added to the steam cracker footprint, 470 kg CO_2e/tonne, giving a total ethylene footprint of 922 kg CO_2e/tonne. Again, this 922 kg CO_2e/tonne is multiplied by the allocation ratio of 1.035 and then added to the polymerization footprint of 328 kg CO_2e/tonne to give a total footprint of 1,283 kg CO_2e/tonne.

Process Footprint (kg CO_2e/T product)	Process Footprint (kg CO_2e/T PE)	Product (Allocation ratio)	Total Footprint (kg CO_2e/T product)	Total Footprint (kg CO_2e/T PE)
275	305	Crude Oil	275	305
		↓ *1.070*		
158	164	Naphtha	452	468
		↓ *1.000*		
470	486	Ethylene *from steam cracking*	922	955
		↓ *1.035*		
328	328	HDPE *(slurry process)*	1283	1283

Figure 4
CALCULATION OF HDPE (SLURRY) FOOTPRINT FROM NAPHTHA, EUROPE

Second, the footprint is shown by tonne of HDPE (in Figure 4, the columns on the second from the left and the far-right). This follows the same logic, except that the footprint is normalized to the final output of polyethylene. Looked at this way – i.e. per tonne of HDPE – the total footprint by main process step is roughly split as follows: 40% to steam cracking, 25% each to crude-oil production and polymerization, with the remaining 10% to naphtha refining.

POLYETHYLENE'S CARBON FOOTPRINT – HOW IT VARIES

Based on the data presented in preceding section, carbon footprints were calculated for polyethylene for the major feedstocks that account for nearly 85% of ethylene production. These were done for several producing regions according to five allocation methods. We applied the four allocation methods defined earlier, i.e. mass all products, economic value all products, mass high-value chemicals and mass prime olefins, and we added another method, economic value high-value chemicals.

Compared to the notional 'base case' of most existing polyethylene footprints – European, naphtha-based production using mass all products allocation – nearly all of the variations examined generate higher footprints. For all three PE types, the downside variation is less than 10%, while the upside ranges around 100% (Table 2).

Table 2
PE'S NOTIONAL BASE CASE IS LOW WITHIN ITS 100% RANGE

	Footprint		
Polyethylene type	*Lowest*	*Notional base case*	*Highest*
HDPE	-7% to base case Middle Eastern ethane/propane, mass all products allocation	European, naphtha-based production using mass all products allocation	+116% to base case Chinese gasoil, mass prime olefins allocation
LDPE	-3% to base case Middle Eastern ethane/propane, mass all products allocation	European, naphtha-based production using mass all products allocation	+119% to base case Chinese gasoil, mass prime olefins allocation
LLDPE	-5% to base case Middle Eastern ethane/propane, mass all products allocation	European, naphtha-based production using mass all products allocation	+85% to base case Chinese gasoil, mass prime olefins allocation

In the following subsections, footprints are presented with respect to the major variables: type of polyethylene, ethylene feedstock, allocation method and region.

Variation by Type of Polyethylene and Polymerization Process

Footprint components – process, fuel and electricity – can vary substantially. Compared to the other types, producing LLDPE can have relatively high process and fuel footprints, while polymerizing LDPE consumes significantly more electricity. The footprint from electricity of course varies by region. The figure below shows the range in polyethylene process footprints for plants located in the USGC. The variations shown in each product type are related to the process technologies used.

Figure 5
VARIATION IN FOOTPRINT BASED ON PE PROCESS, USGC BASIS

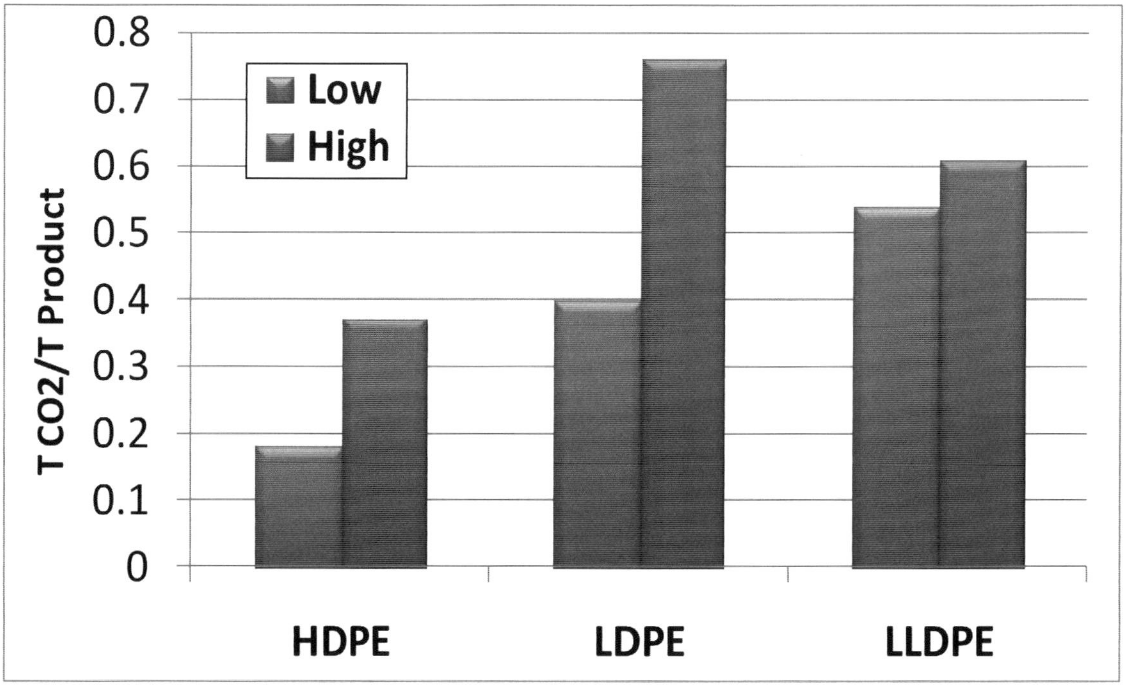

Variation by Ethylene Feedstock

Identifying a consistent ranking of polyethylene footprints according to major feedstock type proves impossible. Without specifying an allocation method, it is impossible to say which feedstock leads to the lowest footprint. As allocation methods change, so do feedstock footprint rankings.

To show this, US footprints were sorted, according to five allocation methods, from highest to lowest by feedstock. Among the five allocation methods, the ranking is different four times. Only with the final two methods (mass HVC and mass prime olefins) does the ranking stay the same.

An attempt was made to boiling this down to a consistent ranking by calculating average footprints (across all allocation methods) and compiling the rankings (Table 3). Still, this does not generate real clarity – the allocation influence is too great.

Table 3
FOOTPRINT RANKINGS, BY FEEDSTOCK

	Number of rankings, out of 5 allocation methods			
Feedstock	Lowest	2nd lowest	3rd lowest	Highest
Atm gasoil		1		4
Naphtha	2		3	
Ethane	3		1	1
Ethane/propane		4	1	

Variation by Region

When regions are compared on the closest to a like-for-like basis, the ranking for energy intensive processes like LDPE are greatly impacted. As shown earlier Table 1 earlier, the CO_2 associated with electricity generation varies widely from country to country due to the variation in fuel mixes used. As shown in Figure 6, this location specific adjustment alone can result in significant variation in footprint of the product in questions.

Figure 6
LDPE FOOTPRINTS, BY REGION
(NAPHTHA, MASS HVC ALLOCATION)

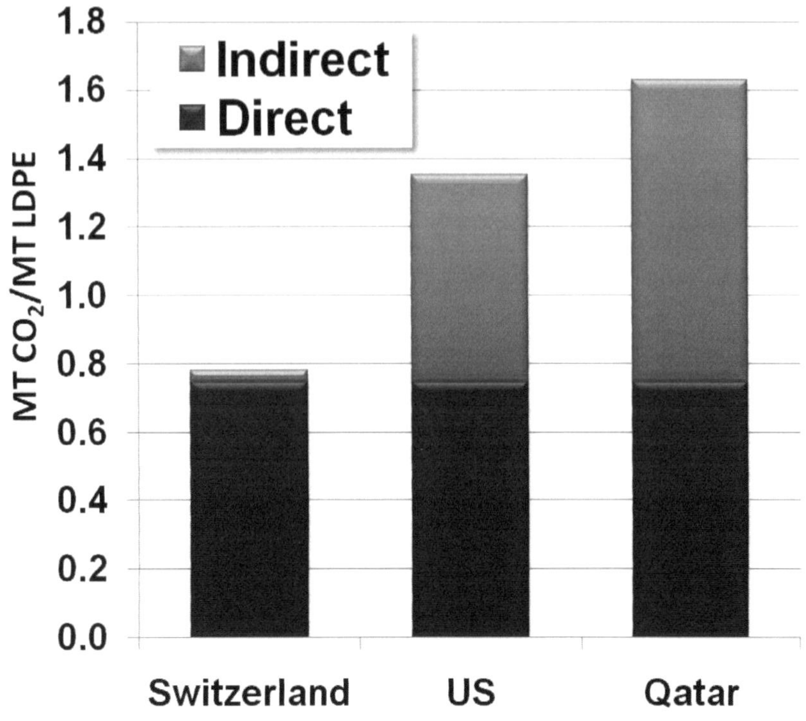

CARBON EMISSIONS FOR BIOETHYLENE FROM SUGAR CANE

Ethanol, currently a minor feedstock for ethylene, could possibly play a much larger role, thanks to increasing interest in biomass as a chemical or fuel feedstock and to the growth of bioethanol. So although it is not yet a commercial reality, as part of this paper we also examined the carbon emissions and footprint of making polyethylene from sugar cane, via bioethylene.

BIO-POLYETHYLENE FROM SUGAR CANE – HOW IT COMPARES

Bio-polyethylene – i.e. polyethylene made from sugar cane – generates a relatively high footprint compared to conventional polyethylene. However, it could be argued that in certain applications and circumstances, bio-PE should be granted a carbon storage credit. In this case, its footprint could drop so far as actually to go negative.

Bio-Polyethylene's Carbon Footprint

Polymerizing bioethylene to polyethylene generates cradle-to-gate footprints roughly comparable to those generated in Europe under the 'mass prime olefins' allocation method.

The difference between conventional and bio polyethylene is all down to the differences in producing the ethylene. Our estimate assumes that bioethylene follows the same route to polyethylene as conventional ethylene does.

Granting a Carbon Storage Credit to Bio-Polyethylene

It could be argued that bio-polyethylene is a means of carbon capture and storage, because it captures carbon (by sugar cane growth) and then stores it, either in long-life products (say, construction materials) or even in landfill, where its degradation rates are very slow. This argument is not at all far-fetched; indeed, it is supported by the primary guideline for carbon footprinting, PAS 2050 (BSI, Carbon Trust et al. 2008).

What would the credit be? Presumably it would be the amount of CO_2 removed from the atmosphere and stored in the polyethylene. Chemically speaking, polyethylene is basically nCH_2; with a carbon weight-fraction of $12/14^{th}$ or 85.7%. Every tonne of polyethylene is thus 857 kg C, which multiplied by 44/12 equals 3,143 kg CO_2e. If this carbon credit is granted, bio-polyethylene's footprint drops dramatically, actually going negative.

Is this result some sort of accounting trick? No more so than the rest of carbon footprinting. It complies with current footprinting rules, it is the same method applied to conventional polyethylene, and at least to the authors, the result makes intuitive sense.

At the same time, the credit would not apply in all polyethylene applications – only to those where the PE exists for a long time. The full credit of 3,143 kg CO_2e per tonne would be granted to PE that lasts for 100 years (the time horizon for global warming calculations), a half-credit to PE lasting 50 years and so on. Similarly, the credit could be granted to the portion of the PE waste that lasted for 100 years (say, a landfilled fraction), while an incinerated fraction would not receive the credit.

Another consideration is the inherent neutrality of the carbon being used in the polyethylene. Footprint accounting's convention – until recently – has been to debit an emission of fossil carbon as soon as it hits the atmosphere, and to net out bio-carbon (i.e. carbon emissions equal carbon up-take). However, this latter method has been challenged by recent studies (Johnson 2009). Our provisional view of the bio-polyethylene from sugar cane case is that a carbon credit could likely be applied, but probably not to 100%.

WHY FOOTPRINT VARIATIONS ARE IMPORTANT

Findings of previous chapters on polyethylene footprints' variability and comparability raise an obvious question: how significant is this variability or comparability?

Our answer: very significant. Footprints are increasingly being used to make purchasing and policy decisions. Differences in footprint measurements can affect those decisions, as is shown by the example below of shopping bags: HDPE vs. LDPE vs. biodegradable vs paper.

Footprints are Being Used in Purchasing and Policy

Because of concerns about global warming, governments, retailers and consumers are demanding carbon footprints for many common products such as cars, homes, household goods, fuels and packaging – such as polyethylene and its competitors. Their aim is to use footprint data to guide policy, commercial and personal decisions.

Carbon footprints are being used by the European Union, some of its member states and the US as guidance in setting fuel taxes. The idea is that lower-carbon fuels should be encouraged through lower taxes, and higher-carbon ones discouraged by higher ones. Carbon footprints are also being used to set vehicle taxes and building codes.

Major retailers such as Office Depot, Tesco and Walmart are starting to offer so-called *carbon labels* on some of their merchandise. These are carbon footprints, printed on the pack. Obviously carbon labels are, at least in part, a response to consumer demand – a personal decision by consumers to measure and lower their footprints. Organizations as diverse as the Barclays Bank, International Football Federation (FIFA), KPMG, Marks & Spencer, Nike, Timberland and Yahoo! claim to have become or that they will become carbon neutral. One of the few quantitative indicators of this commercial/personal trend is the burgeoning market for voluntary carbon offsets. To counterbalance their greenhouse gas emissions, Europeans and North Americans in 2007 spent as much as $200 million on voluntary offsets and by 2010 they may spend up to $4 thousand million (World Bank 2007).

WHAT THIS MEANS FOR POLYETHYLENE PRODUCERS AND USERS

Based on the work presented in the previous chapters, the authors come to the following conclusions for those in the polyethylene industry:

- Footprint competition is possible
- You should start measuring your footprint – and there is a cheap, easy way to do this
- Allocation probably can and certainly should be sorted out

Footprint Competition: Middle East Looks Good, China Not So Much

There is a broad variation in polyethylene footprints, with a range of nearly 3:1 at the extreme. These differences can be used positively or negatively in marketing and promotion, within the industry or for or against other industries.

For instance, producers using ethane/propane feed in the Middle East – arguably the marginal source of production – might point out that their polyethylene could have about one-third the footprint of polyethylene produced in China. They also might contend that shopping bags made from their polyethylene beat paper ones, while the Chinese ones do not.

Start Estimating Your Footprint – Easily and Cheaply

To act or react in footprint competition, players throughout the polyethylene chain will want to know their footprints. An easy, inexpensive way to assess this is by using the footprints published by third-party sources, such as SRI Consulting's Greenhouse Gas Handbook. After an initial assessment, it can be determined then whether more data-gathering or analysis is needed.

Which Allocation Method is Best?

The method chosen to allocate a steam-cracker's footprint is critical to the polyethylene footprint, and using different, accepted methods can result in major differences – so major that they can reverse conclusions drawn from them. Ideally, one allocation method should be specified and applied to ethylene footprints, one that would eliminate these differences.

Probably the best method would be one that:

- Generates the least 'wobble' with respect to decision criteria, i.e. technology, feedstock, product type or region. By wobble, we mean varying answers that stymie consistent conclusions. (The wobble factor is a common criticism of economic allocation methods.)
- Is not too difficult to compile
- Is believed to be credible by users of footprints.
- Reflects underlying economic incentives.

On this basis, we would propose that the standard allocation method should be mass high value chemicals (HVC). This method:

- Appears to generate minimum wobble
- Is far less difficult to compile than an economic allocation, yet by restricting itself to HVCs, it is to some extent an economic allocation anyway.
- Is credible, in that the IEA has been using it in petrochemical studies
- Reflects economic incentives, i.e. it allocates footprint only to value-added products.

We recommend that ISO and BSI, the authorities behind existing allocation standards, consider adopting mass HVC as a standard method.

References

1. Argonne Labs GREET GREET, Greenhouse Gases, Regulated Emissions, and Energy Use in Transportation, Version 1.8c.
2. BSI, Carbon Trust, et al. (2008). PAS 2050. Specification for the assessment of the life cycle greenhouse gas emissions of goods and services. Publicly available specification 2050:2008.
3. Ecobilan (2004). Evaluation des impacts environnementaux des sacs de caisse Carrefour.
4. EcoInvent LCI Database. St Gallen, Switzerland. Current database is V 2.0.
5. EcoInvent (2007). Plastics: LCIs of packaging and graphical papers. EcoInvent Report No. 11.
6. IPCC (2006). IPCC Guidelines for National Greenhouse Gas Inventories.
7. ISO (2006). ISO 14040: Environmental management — Life cycle assessment — Principles and framework.
8. ISO (2006). ISO 14044: Environmental management — Life cycle assessment — Requirements and guidelines
9. Johnson, E. (2009). "Goodbye to carbon neutral: getting biomass footprints right." Environmental Impact Assessment Review.
10. Joint Research Centre of the EU Commission, EUCAR, et al. (2006). Well-to-Wheels analysis of future automotive fuels and powertrains in the European context.
11. Patel, M., M. Neelis, et al. (2005). "Carbon dioxide emissions from non-energy use of fossil fuels: Summary of key issues and conclusions from the country analyses." Resources, Conservation and Recycling 45 (2005) 195-209.
12. Patel, M. and D. Saygin (2008). (Footprints of) Chemical and Petrochemical Sector, International Energy Agency.
13. Plastics Europe (2005). Crude Oil. Eco-profiles of the European Plastics Industry. Brussels, Plastics Europe.
14. Plastics Europe (2005). Ethylene. Eco-profiles of the European Plastics Industry. Brussels, Plastics Europe.
15. Plastics Europe (2005). Naphtha. Eco-profiles of the European Plastics Industry. Brussels, Plastics Europe.
16. Plastics Europe (2005). Natural gas. Eco-profiles of the European Plastics Industry. Brussels, Plastics Europe.
17. SRI Consulting (2007). Carbon Footprints of Biofuels & Petrofuels. Greenhouse Gases.
18. SRI Consulting (2007). Greenhouse Gases Handbook. M. Arne.
19. SRI Consulting (2008). Ethylene Carbon Footprints, Will the Real Carbon Footprint Please Stand Up? Sensitivities of Ethylene Carbon Footprints to Feedstock Type, Allocation, Method and Location. PEP Review 2008-12.
20. SRI Consulting (2009). Polyethylene's Carbon Footprint:Global Variations And Comparisons.
21. World Bank (2007). State and trends of the carbon market 2007, World Bank.

AIChE Paper Number 12b

PROGRESSION OF GLOBAL GHG EMISSIONS BENCHMARKING

Claire L. Cagnolatti

Vice President, Chemicals Studies

HSB Solomon Associates LLC

Kevin Henke

Senior Consultant

HSB Solomon Associates LLC

Prepared for Presentation at the 2010 Spring National Meeting
San Antonio, TX, March, 21-25, 2010

PROGRESSION OF GLOBAL GHG EMISSIONS BENCHMARKING

Claire L. Cagnolatti

Vice President, Chemicals Studies

HSB Solomon Associates LLC

Kevin Henke

Senior Consultant

HSB Solomon Associates LLC

Abstract:

The efforts to measure, regulate, and control the emissions of greenhouse gases (GHG) by the refining and petrochemicals industries have been in effect in varying degrees in several countries and regions around the world. Companies that operate such facilities in regions with pending or active regulation of GHG have sought the assistance of Solomon, a consulting firm that has conducted benchmarking studies for these industries since the 1980s. In this paper, Solomon shares their experiences gained from working with companies, industry associations, and regulatory agencies around the globe. Topics covered include following:

- Methodologies used and accepted by regulatory agencies
- Cooperation between industry associations and regulatory agencies
- Allocations vs cap-and-trade methods of limiting emissions
- Intensity-based and complexity-based benchmarks, such as the Carbon Emissions Index (CEI™) metric vs emissions per unit of production or throughput

Examples of timelines for regulatory negotiations and supporting analyses, as well as experiences gained in these negotiations, will also be discussed.

Introduction

The efforts to measure, regulate, and control greenhouse gases (GHG) emissions by the refining and petrochemicals industries have been in effect in varying degrees in several countries and regions around the world. Companies that operate such facilities in regions with pending or active GHG regulations have sought the assistance of HSB Solomon Associates LLC (Solomon), a consulting firm that has conducted Comparative Performance Analysis™ (CPA™) benchmarking studies for these industries since the 1980s.

Solomon's History with GHG Measurement (Voluntary Arena)

Several governmental authorities have adopted some of Solomon's standard measures to drive energy and emissions efficiency. Energy Intensity Index (EII®), for example, was developed for Solomon's *Fuels Refinery Performance Analysis (Fuels Studies)* in the early 1980s and later (2008) adapted for ethylene producers in the *Worldwide Olefin Plant Performance Analysis (Olefin Study)*. This metric compares the Solomon-calculated standard energy consumption to actual energy consumption. Programs now using EII include the U.S. Environmental Protection Agency ENERGY STAR program, American Petroleum Institute, and the Netherlands Government's Energy Covenant program.

Applying a similar methodology as EII, Solomon developed a measure of GHG emissions efficiency—the Carbon Emissions Index (CEI™). Solomon calculates standard carbon dioxide (CO_2) equivalent emissions (similar method to standard energy) and compares that value to actual CO_2 equivalent emissions. CEI was established in 2006 for refineries and olefin plants and has been included in our study deliverables since that time. Its use as a management tool within our client companies is growing, but due to its rigorous calculations and proprietary nature, CEI was not suitable for regulatory purposes.

Involvement in the Regulatory Arena

To meet the need for metrics with the transparency and relative simplicity required for regulatory purposes, Solomon developed customized energy and/or emissions intensity measures. We have worked with several manufacturing associations around the world, in conjunction with and on behalf of their member manufacturing companies who are also our clients:

- Refining – Petroleum Association of Japan (PAJ), Canadian Petroleum Products Institute (CPPI), and Western States Petroleum Association (WSPA) in California
- Olefins – European Chemical Industry Council (CEFIC) and the Association of Petrochemical Producers of Europe (APPE)

- Customized Allocation Methodologies
  - Refining – UK Department of Trade & Industries (DTI)
  - Refining – EUROPIA/CONCAWE in Europe

Key Points

As one would expect, Solomon has uncovered some key points and general principles through our experiences in helping clients and their manufacturing organizations. These key points are universal for refineries, olefin plants, or any other manufacturing entity.

Tradeoff between Simplicity and Fairness

Regulatory agencies have a strong need and desire for transparency and openness in emissions measurement methodology, with a simple basis that is easy to understand, communicate, and regulate. Additionally, manufacturers whose emissions are to be measured have a strong need and desire for fairness. Unfortunately, the fairest measurement methods are usually the most complex. So, a balance must be found between the simplicity of a benchmarking method and the equity in distributing the burden of compliance. That balance must not sacrifice equity for the sake of simplicity.

To achieve simplicity in emissions measurements, a method should use the fewest variables from easily obtained data, without sacrificing accuracy. A simple method also should be suitable for consistent application across the industry via homogeneous rules, with little or no need for "special" exceptions or corrections. Regulators require that the chosen measurement method be open and transparent, using no inaccessible or proprietary data or methods.

An equitable measurement must differentiate performance among manufacturing plants in a reliable and predictable manner, including performance differences due to varying unit complexities. It must also give credit to those who have taken early action and not reward the lack of action. This approach ensures that those who have already achieved significant improvement are not expected to have the same magnitude of improvements as those who have yet to take action.

Other Objectives

An optimal emissions measurement method should promote or enable gains in energy efficiency, reductions in hydrocarbon flaring, switching to minimum-emissions fuel types, and the employment of carbon-efficient technologies. It should also be intensity-based to allow fair allocations of emissions—higher allowances for processes with higher energy intensity—as a regulatory goal.

Finding the Right Balance

Facilities vary significantly from refinery to refinery, so a simple emissions benchmark such as CO_2 per ton or barrel of throughput would not meet the criteria of fairness and ability to differentiate by complexity. Therefore, Solomon worked with refining clients to develop the complexity-weighted metric ton (CWT™) metric based on CO_2 intensity. The complexity-weighted barrel (CWB™) metric is the volume-based version of CWT. These complexity measures take into account the sum of the complexities of the units that make up any individual refinery.

With regard to olefin manufacturing, Solomon has worked extensively with European producers via the manufacturing organizations CEFIC and APPE to find the optimal method to meet their needs. Within Europe, olefin plants crack very similar feeds, so there was less need for differentiation due to plant configuration. The goal was to achieve the most equitable treatment with a relatively simple and transparent method.

Solomon's Work with European Ethylene Producers

Following are some of the issues and concerns raised during the process of developing an equitable emissions measure for European ethylene producers:

- Emissions benchmarking could be a hindrance to regional competitiveness.
- The long-term viability of the industry must be maintained, and the economic impact of any regulation must be quantified to be fully understood.
- Possibility of "carbon leakage" (CO_2 emissions being displaced from highly-regulated regions to minimally-regulated regions) must be taken into account.
- Benchmarking (measurement) must be suitable and fair.
- Allocation/limit method must be suitable and fair.
- Benchmarking method must be transparent.
- What sensitive operating data will be made public?

Solomon's work with CEFIC began in 2007 with the review of an International Energy Agency (IEA) report on energy consumption and emissions from petrochemicals manufacturing, in which Solomon called attention to some over-simplifications, misinterpretations, and omissions. From 2007 to the end of 2009, Solomon worked with European ethylene producers to test various bases for emissions (direct fuel only, fuel plus import steam, fuel plus import steam plus import electricity); various inclusion boundaries (cracker only, cracker plus other petrochemical units); and using both feed and high-value chemical (HVC) products as divisors. Solomon also assisted in the development of the input forms to quickly and accurately collect the emissions data.

The method and benchmark chosen by the European ethylene producers will be presented to European Union authorities in February 2010, which is after the conference deadline for the text of this paper. After the information is presented to the European authorities, it will be essentially in the public domain, and will be covered in this paper's Microsoft® Power Point presentation made at this conference.

AIChE Paper Number 12c

CARBON CAPTURE IN CRACKING FURNACES

Markus C. Weikl

Process Development

Linde AG

Gunther Schmidt

Manager or Process Engineering – Cracking Furnaces

Linde AG

Prepared for Presentation at the 2010 Spring National Meeting
San Antonio, TX, March, 21-25, 2010

CARBON CAPTURE IN CRACKING FURNACES

Markus C. Weikl

Process Development

Linde AG

Gunther Schmidt

Manager of Process Engineering – Cracking Furnaces

Linde AG

Abstract:

Carbon dioxide (CO_2) capture and storage (CCS) will be an important part of the greenhouse gas mitigation portfolio if ambitious targets as stabilization of atmospheric CO_2 concentration at a level of 450 ppm should be met in the future. Besides the energy sector, other CO_2 emission intense sectors as fuel processing, iron & steel, cement, pulp and paper, and chemicals production will also have to contribute by applying capture schemes.

This paper presents a techno-economic study of carbon capture from an ethylene production plant. First, the main trends of carbon capture in the power plant sector and the boundary conditions for a transition of these techniques to the chemical sector, i.e. an ethylene plant, will be discussed.

The boundary conditions regarding type of feed and product related CO_2 emissions are presented. The applicability of capture technologies is discussed and capture of CO_2 by the oxyfuel- and post-combustion process is investigated.

The application of oxyfuel-combustion allows for increased efficiency in the radiant section compared to air/fuel combustion and therefore gives additional benefit for the use of this type of technology. In contrast, post-combustion capture can be regarded as an add-on solution with fewer changes to the cracking furnace, if low pressure steam for regeneration of the solvent is available. A case study for CO_2 capture from a set of five cracking furnaces with 100 MW firing duty each is conducted.

The resource and utility requirements for both capture technologies are evaluated and compared to a base case scenario under CO_2 emission trading. Furthermore, capture costs are calculated and the preferential choice of capture technology is discussed relative to the carbon prices in an emission trading scheme.

Introduction

It is nowadays widely accepted that climate change is a major challenge and is caused mainly by anthropogenic emissions of greenhouse gases. Carbon dioxide capture and storage (CCS) will be an important part of the greenhouse gas mitigation portfolio if ambitious targets as stabilization of atmospheric CO_2 concentration at a level of 450 ppm or a maximum of 2°C global warming related to preindustrial times should be met. The International Energy Agency (IEA) Technology Roadmap for Carbon Capture and Storage[1] estimates that by 2050 approximately 20% of the emission reductions will need to be achieved by implementation of CCS. A major share of this emission reduction is of course to be realized by the energy sector. However, CCS is more than a strategy for clean coal and must be adopted also by other emissions-intensive industrial sectors like cement, iron and steel, chemicals, and pulp and paper. It is stated that by 2050 roughly one third of the required CCS deployed capacity is to be situated in the industrial sector. The distribution among sectors, shown in Figure 1, is a function of capture costs because emission reduction by CCS will be realized first in a sector, if they have lower costs compared to CCS in other sectors and emission trading.

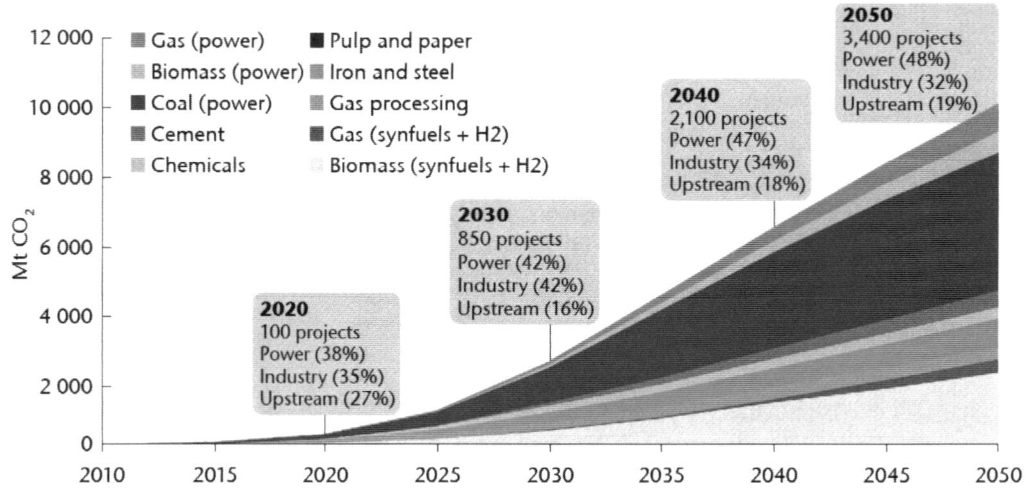

Figure 1: Global deployment of CCS by sector.[1]

For this sectors, viability of CCS and cost range need to be assessed. Early investigation of this topic was conducted by Farla et. al.[2]. However, also in newer literature no specific investigation of carbon capture from cracking furnaces can be found.

This paper aims at the techno-economic investigation of carbon dioxide capture from the furnace section of an ethylene production plant. Firstly, a short review of carbon dioxide capture in the energy sector will be conducted.

CCS in the energy sector

Currently, three main branches of capture technologies are considered for power plants, see Figure 2. Due to the high specific CO_2-emissions of at best 700-800 g/kWh mainly coal-fired power plants will be equipped with these capture processes.

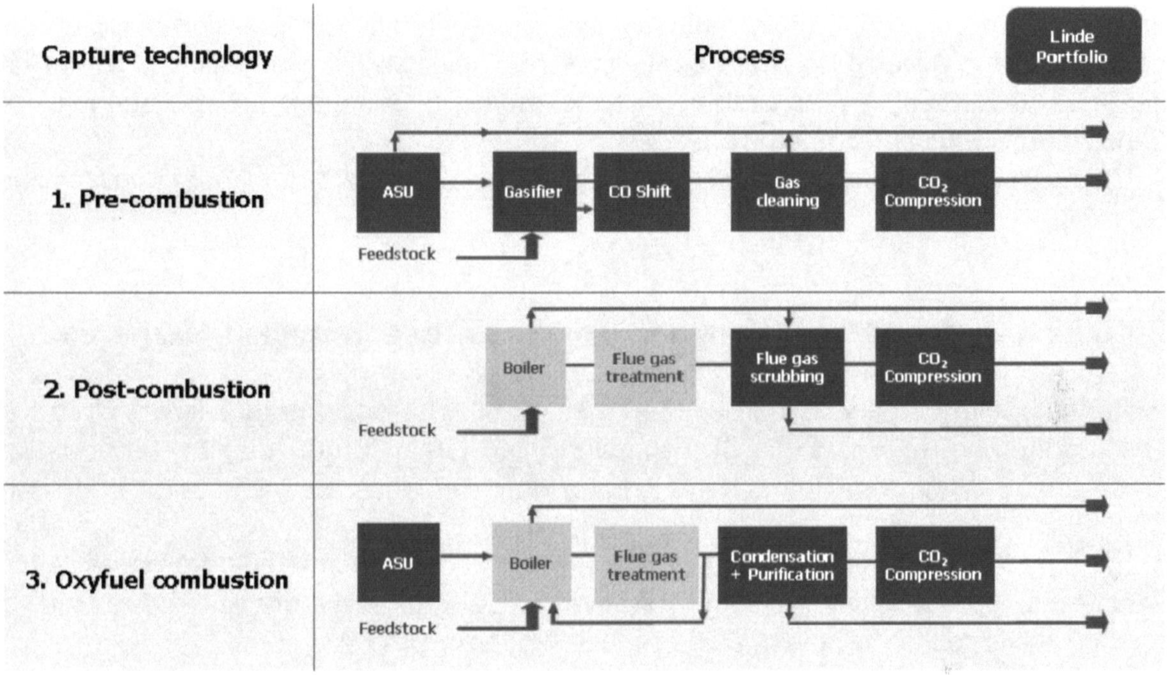

Figure 2: Carbon capture processes considered in the energy sector.

Pre-Combustion capture can be implemented in form of an integrated gasification combined cycle (IGCC) process. First, in a gasification process coal is converted into synthesis gas, which is subsequently treated in a CO-shift conversion to increase the hydrogen content. The shifted gas is treated by a chemical or physical absorption unit, i.e. Rectisol®, to extract CO_2 and leave a hydrogen rich fuel gas for a combined cycle power generating process. At present, development of hydrogen driven gas turbine systems is the main focus of development. The process benefits from a high efficiency in the combined cycle process and smaller unit sizes compared to pulverized coal power plants due to process pressures of 10-60 bar.

Post-combustion capture is characterized by flue gas scrubbing with a chemically active solvent, which reacts with the CO_2 at atmospheric pressure. The parasitic energy loss is in this case mainly attributed to the use of steam, which is extracted from the low pressure turbine, for regeneration of the solvent. The baseline process is employing Monoethanolamine (MEA) as a solvent. Currently, the aim is to develop new solvents which allow for more effective separation of CO_2. The ideal solvent shows quick reaction kinetics and high cyclic capacity and low demand for regeneration energy as well as low degradation.

Oxyfuel combustion involves the substitution of air sent to the pulverized coal burner with pure oxygen. Here, the benefit is that after cooling and condensation of water a stream which consists of a high concentration of carbon dioxide is received. This stream can be cleaned comparatively easy to a high purity stream for sequestration in geological formations. First experiences at pilot plants, e.g. Vattenfall's Schwarze Pumpe[3] , show viability of the process.

All of the mentioned processes have in common, that by parasitic power consumption of the capture unit, the overall power plant efficiency of converting primary energy into electricity drops by about 10%. Parasitic power consumption involves e.g. separation of oxygen from air, regeneration of solvent, purification, drying and compression of CO_2-product. The next phase for developing these capture technologies with demonstration plants having thermal sizes of 200-600 MW is prepared at the moment. Commercial deployment of power plants including CCS is anticipated from 2020.

Transport and storage of captured CO_2 are basically independent from capture processes.

Boundary conditions for carbon dioxide capture in cracking furnaces

The ethylene industry as one of the biggest bulk chemicals producer will have to deal with the requirement to reduce its CO_2 emission in the near future. Due to the endothermic nature of the cracking reaction a high combustion duty has to be fired in the cracking furnaces.

Compared with the energy sector there are, however, remarkable differences. Although the plants have been steadily growing and have reached an accumulated fired duty of up to almost 1000 MW_{th} coal fired boilers are yet much bigger in this respect.

The fuels as such are also different. Steamcrackers are producing a tail gas or methane fraction which is preferably burnt in the cracking furnaces to supply the heat necessary for the endothermic cracking reaction. Since these gas fractions are usually containing only very low concentrations of corrosive elements (e.g. sulfur) the heat recovery is already fairly high at typically approx. 93-94 %. Consequently, saving CO_2 by increasing the thermal efficiency, gives only minor potential. Typical fuel gas compositions and the specific CO_2 productions are shown for different feed stocks in Fig. 3. It can easily be taken that any carbon capture consideration should concentrate on those plants firing fuels with low hydrogen content, namely on liquid crackers.

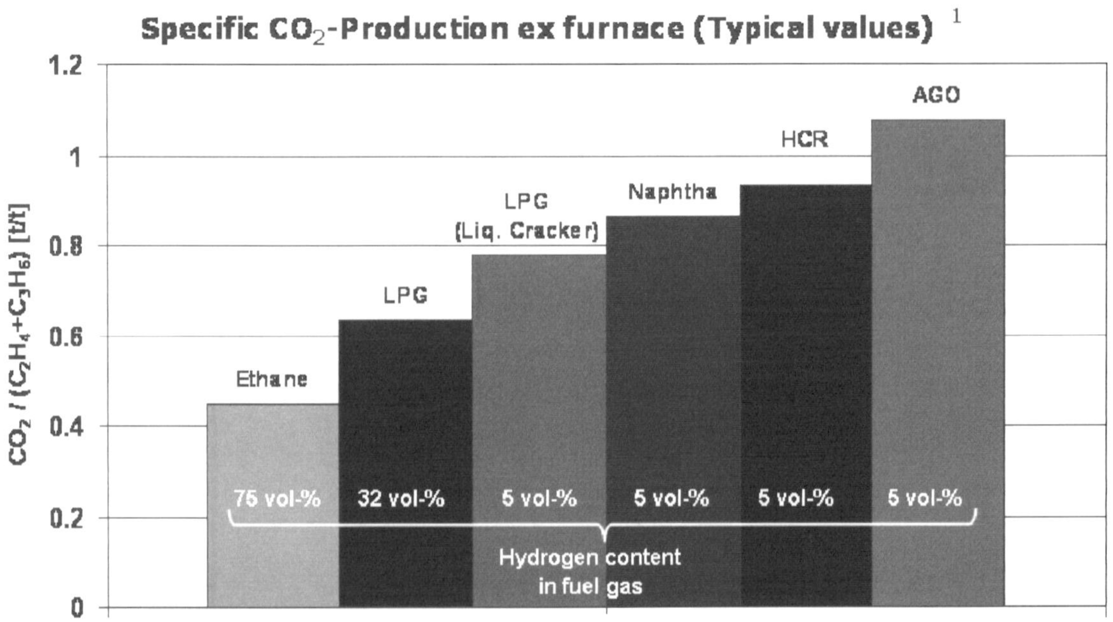

Note 1: Shown figures refer to expected direct ethylene and propylene production in cracked gas and CO_2 production in flue gas ex furnace. Figures based on different cracking severities, furnace designs and sizes.

Figure 3: Specific CO_2-Production ex furnace

Options for cracking furnaces

Post combustion treatment:

Post-combustion treatment can be similarly applied to cracking heaters as to any combustion system. It is typically regarded as an add-on solution with low changes to the cracking furnace itself. A stronger induced draft fan is needed to overcome the pressure drop in the absorber. For regeneration of the amine solvent low pressure saturated steam of 4 bara is required. The typical flue gas composition of an air/fuel fired furnace contains about 9% CO_2 compared to 12-15% in a pulverized coal boiler. Hence, regeneration energy will be slightly higher in the case of a cracking furnace. Typical values for the regeneration energy of MEA are within 3.4-4.0 GJ/t_{CO2}. For this work 3.8 GJ/t_{CO2} are assumed. After regeneration, the CO_2 stream with a purity of >98% is compressed and dried. Regarding quality requirements, it is estimated, that an oxygen level of 200 ppm is present in the CO_2 product after compression and drying. Hence, without further measures it is not possible to use this stream for EOR and a direct comparison to the oxyfuel cases applies only for sequestration in saline aquifers (see below).

Oxyfuel combustion

There are two different application modes when pure oxygen is to be fired:

a) with internal flue gas recycle
b) with external flue gas recycle

For both, the basic key criteria is to ensure that an evenly distributed heat flux to the cracking coils is maintained. Furthermore, maximum temperatures at the tube walls need to be limited according to metallurgic requirements. Combustion with internal recycle, i.e. mixing after the burner tips, requires a completely new burner and furnace design, to be able to limit the adiabatic combustion temperature to acceptable levels. Hence, a mid-term deployable furnace with oxyfuel combustion would employ external recycle of flue gas. Due to basically zero SO_x concentration, the recycle can be taken directly after the convection section, which reduces stack losses and increases thermal efficiency. For comparison, in coal fired power plants the recycle will due to massive corrosion problems in the future probably be taken after the flue gas desulphurization unit.

In any case oxyfuel combustion will affect the heat balance. Due to the radiation properties of CO_2 and H_2O the heat transfer in the radiant section will be more efficient by about 10%. This leads to reduced firing duty but on the other hand reduces high pressure steam export

The initial and modified heat balances are shown in Figure 4.

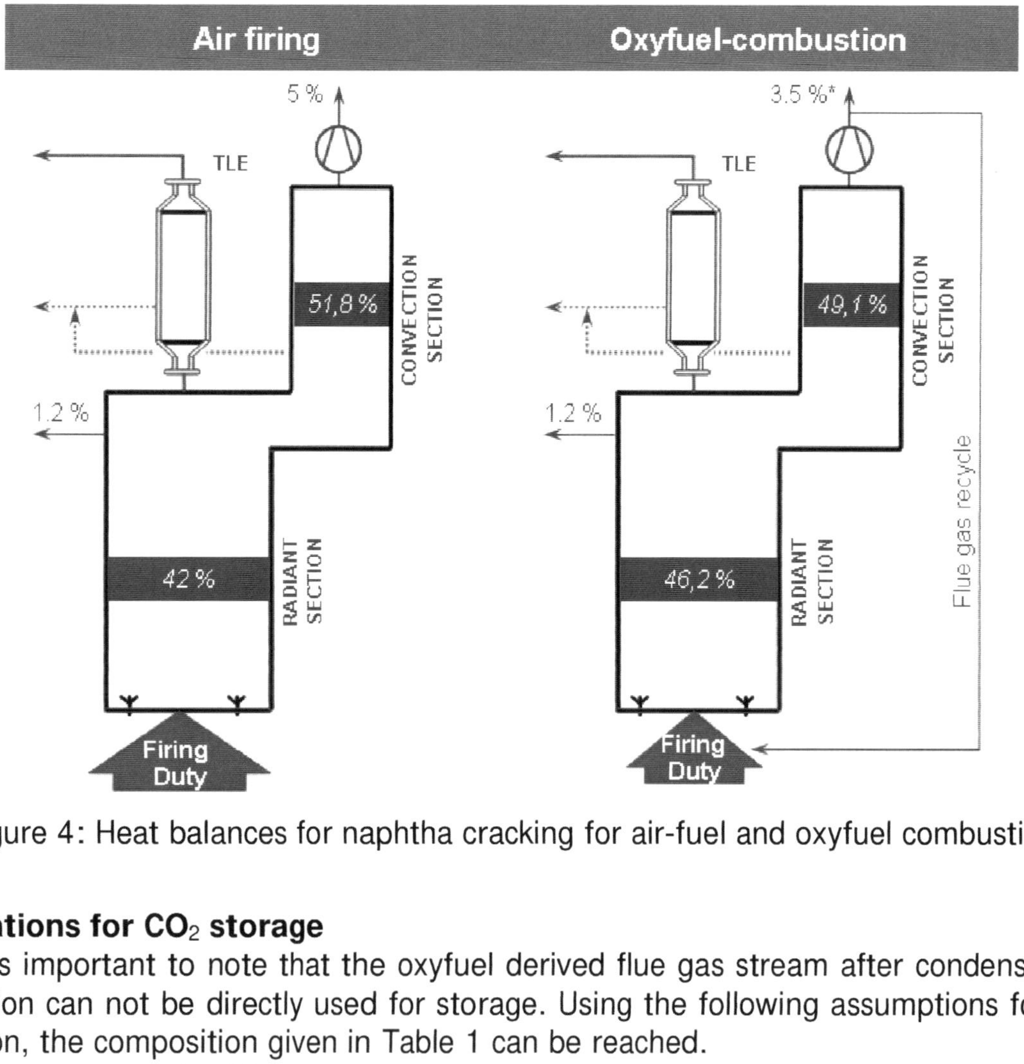

Figure 4: Heat balances for naphtha cracking for air-fuel and oxyfuel combustion

Specifications for CO_2 storage

It is important to note that the oxyfuel derived flue gas stream after condensation and compression can not be directly used for storage. Using the following assumptions for oxyfuel combustion, the composition given in Table 1 can be reached.

- oxygen purity 95%
- excess oxygen to the furnace: 10%

- flue gas recycle ratio (FGR): 70%
- condensation of water
- water knock-out during compression and intercooling up to 100 bar
- no air leakage to the flue gas

Table 1 compares the requirements for the enhanced oil recovery (EOR) project at Weyburn and specifications from the Dynamis project for storage in Saline Aquifers.[4]
It is shown that the flue gas from oxyfuel combustion needs to be treated in a purification plant by liquefaction and separation of non-condensable gases to the required levels. Two cases were investigated in analogy to two CO_2 quality requirements from Table 1, i.e. Weyburn for EOR and storage in saline aquifers according to Dynamis specifications.[4] For Dynamis, non-condensable gases are to be limited to a level of below 4%, including oxygen. In EOR, oxygen is a possible reactant with the oil and may lead to clogging in the reservoir. Hence, quality requirements regarding oxygen impurities are more stringent and imply treatment of the liquefied CO_2 by rectification. For saline aquifer storage a maximum O_2 level of <1% and for EOR a level of <50 ppm was assumed.

Table 1: Composition of oxyfuel derived flue gas after condensation and compression and requirements for storage (NCG: non condensable gases).

	oxyfuel flue gas	Weyburn[4]	Dynamis[4]
CO_2	85%	>96%	>95.5%
CO	dep. comb.	<1000 ppm	<200 ppm
Total HC	dep. comb.	<3%	<2-4%
H_2O	1000 ppm	<20 ppm	<500 ppm
H_2S (SO_x)	dep. fuel	<9000 ppm	<200 ppm
Inerts (N_2+Ar)	9.4%	<1300 ppm	<4% NCG
O_2	5.5%	<50 ppm	incl. in NCG

In general, for a real storage project, the quality requirements will be assessed specifically, since there are up to now no general regulations or laws available.

Case study

A case study was conducted, which assumes the following scenarios:

- Conventional air-firing without capture but with emission trading
- Air-firing with post-combustion capture for saline aquifer storage
- Oxyfuel combustion capture with quality requirement for
  - saline aquifer storage (SA)
  - enhanced oil recovery (EOR)

To investigate a reasonable size and apply investment data for the capture plants known from the energy sector, a set of five cracking furnaces with a total production of 80 t/h of ethylene product has been selected. The results of specific consumptions and important streams are given in Table 2. Due to the higher efficiency in the radiant section in the oxyfuel case firing duty is 100 MW per furnace whereas it is increased by 10% in the air-fuel cases. Consequently, overall CO_2 produced, which is a sum of CO_2 emitted and CO_2 captured is higher for the air-fuel cases. For the cases with carbon capture a capture rate of 90% was chosen. It has to be noted, that by increased investment in oxyfuel operation a capture rate of up to 99% can be achieved, e.g. by an adsorption unit for CO_2 from the vent gas. The amount of CO_2 emitted has a negative effect in plant economics under a cap-and-trade system and is the main driver for establishment of a capture unit. The value of certificate costs is a variable in the cost model. The capture units have electrical demand on the one hand for oxygen generation and on the other hand for compression and purification. Compared to coal-fired power generation, specific demand for oxygen generation is approximately by a factor of 2 higher due to the lower carbon intensity of the fuel.

Table 3: Comparison of specific consumptions and streams.

		Air-fuel No Capture	Post-Comb. SA	Oxyfuel SA	Oxyfuel EOR
Firing duty	%	+10	+10	base	base
CO_2 captured	t/h	0	96	87	87
CO_2 emitted	t/h	107	11	10	10
O_2	kWh/t_{CO2}	0	0	296	296
Compression (+Purification)	kWh/t_{CO2}	0	121	154	164
Steam demand	t/h	0	180 (LP)	38 (HP)	38 (HP)

Energy demand for compression is lowest for post-combustion capture due to a higher initial purity and no requirement for liquefaction to reduce the content of non condensable gases. For EOR-quality, power requirement are higher due to liquefaction and rectification in the CO2-plant.

Low pressure steam demand for Post-combustion is comparatively high due to the need for regeneration of the solvent. For the oxyfuel cases, net high pressure steam export is reduced, which needs to be considered in the cost model.

In the cost model, prices for fuel, steam (LP, HP), electricity and CO_2 emission certificates and investment costs are set to infer the additional ethylene costs and the brake even point for CO_2 capture in comparison to the emission scenario. Annuity method with an interest rate of 8% and 15 years plant lifetime was used. For operation and maintenance an annual value of 5% of investment costs was applied.

Conclusion

It is necessary to apply a penalty (certificate price) on CO_2 emission of approx. 85 \$/ton (60 €/ton) to arrive at a break-even point. Hence, the additional ethylene costs are on the level of 127 \$/ton (90 €/ton) Ethylene.

The capture costs are remarkably higher compared to CO_2 capture in coal fired power plants, due to the low carbon intensity of the fuel and therefore the comparatively high value of specific oxygen demand.

The thermal efficiency increase by oxyfuel firing compared to standard air firing is only marginal and, thus, there is no advantage of oxyfuel firing over post-combustion separation.

The utilisation of CO_2 for EOR is roughly able to pay the capture costs if an additional oil recovery of 2 bbl/t_{CO2} is assumed at an average oil price of 70 \$ (50€) per barrel.

It has to be noted that there are big uncertainties regarding investment cost estimates and regarding the efficiency of improved future post-combustion processes. Closer feasibility investigations have to consider specific boundary conditions respective to:

a) steam costs and steam availability for selection of the capture process
b) specification for CO_2 depending on further usage
c) costs for transportation and storage
d) plant life time, annuity, variable costs (power, steam, ...)

Last not least, it should be mentioned that the additional power and steam demands lead to additional CO_2 generated compared to the pure cracking process by 20-30%.

References

[1] International Energy Agency (IEA): Technology Roadmap Carbon Capture and Storage, IEA/OECD 2009.
[2] J.C.M. Farla, et. al.: Carbon dioxide recovery from industrial processes, Climatic change 29: 436-461, 1995.
[3] H. Altmann et. al.: Experiences from the Commissioning and the Trial Operation of the Oxyfuel Research Plant of Vattenfall, VGB PowerTech 9/2009.
[4] E. deVisser et. al.: DYNAMIS CO_2 quality recommendations, International Journal of Greenhouse Gas Control 2: 478-484, 2008.

AIChE PAPER NUMBER 12d

CARBON CAPTURE AND STORAGE

CAN CCS BE A VIABLE BUSINESS?

Ray P. Hattenbach

Vice President

Blue Source LLC

Prepared for Presentation at the 2010 Spring National Meeting

San Antonio, Texas, March 22-25, 2010

CARBON CAPTURE AND STORAGE

CAN CCS BE A VIABLE BUSINESS?

Ray P. Hattenbach

Vice President

Blue Source LLC

Abstract:

Carbon capture and storage (CCS) has been characterized as the technology that can jump start mitigation of greenhouse gases until new, less carbon intensive technology can be implemented to provide electrical power, transportation fuels and energy our society needs to stay competitive in the world economy. This paper will review the current legislation that has been proposed by Congress, the House of Representatives and the Senate, the effect the proposed regulations will have on existing CCS operations and the impact the proposed regulations will have on companies focused on developing CCS into an ongoing economically viable business. Individual states have preempted the Federal government in developing regulations governing the CCS within their boundaries which will also impact the development and build-out of a viable CCS infrastructure.

Today, we have two markets for the CO2 molecule. First and foremost it can be described as a commodity for use in enhanced oil recovery (EOR) where it has been used for almost 40 years in the Permian Basin of Texas. CO2 for EOR has been a viable, established and ongoing business model for several companies but faces potential challenges from pending legislation and future regulations by the EPA. Current infrastructure will be discussed and new areas under consideration for expanded EOR operations will be high-lighted.

Secondly, CO2 has been characterized as an "air pollutant" by the U.S. Supreme Court under the Clean Air Act. This ruling creates the potential for major differences in regulatory oversight and will have major consequences in development of a viable economic business model for the sequestration of CO2 outside of its use for EOR. Economic opportunities for geological sequestration are reviewed with respect to the impact pending legislation could have on this emerging business model. Pending legislation can either jump start or create barriers to entry so high that companies will not get involved because of the perceived risks and long term liabilities that accompany sequestration.

Introduction

Carbon capture and storage (CCS) has been characterized as the technology that can jump start mitigation of greenhouse gases until new, less carbon intensive technology can be implemented to provide electrical power, transportation fuels and other forms of energy our society needs to stay competitive in the world economy. How we meet the growing energy demands while reducing the impact of increased energy usage on the environment is the role that many expect CCS to fill. In the United States, coal which has the highest carbon intensity currently provides approximately 50% of the electric power and will continue to hold that position for the next 25 plus years. There is expected to be a movement to increase fuel switching and implementation of new energy efficiency technologies, but coal will continue to be a dominant energy source in the US. For any appreciable reduction of carbon emissions to be achieved, CCS certainly will play an important role until fuel switching to natural gas and nuclear can make substantial inroads in base load power generation. Even after this 25 plus year transition period, coal will still be responsible for a substantial portion of base load power generation.

US Emissions

The US Energy Information Agency's (EIA) "Annual Energy Outlook 2010" (AEO2010) forecasts that fossil fuels share of energy consumption goes from 84 percent of the US total in 2008 down to 78 percent in 2035. Even with the substantial increase in renewable energy generation, total coal consumption is forecasted to increase from 1,122 million short tons in 2008 to 1,319 million short tons in 2035. EIA's AEO2010 report estimates that power generated from coal will grow by 24 gigawatts from 313 gigawatts in 2008 to 337 gigawatts in 2035 (Figure 1). Even with the increase in renewable power generation, EIA forecasts energy related carbon dioxide emissions will increase 8.7 percent by 2035 from 5,814 million metric tons in 2008 to 6,320 million metric tons in 2035 for an annual increase of about 0.3percent per year (Figure 2). These CO2 emissions are based on the quantity of fossil fuels EIA is estimating will be consumed to generate the majority of electrical power in the US. Coal by far is the largest single contributor to CO2 emissions due to its high carbon content and that it is used to generate about 50 percent of the electric power consumed in the US. The target for CO2 capture is great and will continue to increase for the next 25 years. From today's actual CO2 emissions and the future emissions that are forecast by EIA to be emitted, there is no doubt that there is a potential market for CCS. The dilemma that one faces in being able to move forward aggressively in developing CCS as a viable business is the value that society currently places on carbon and the driving force or lack thereof, to reduce carbon emissions, whether in a regulatory or voluntary framework. Without a substantial increase in the economic value of carbon reductions, no significant advances will be made in CCS due to the high capture cost with today's proven technology.

Figure 1. Electricity Generation by Fuel: 1990-2035

Renewables gain electricity market share; coal share declines

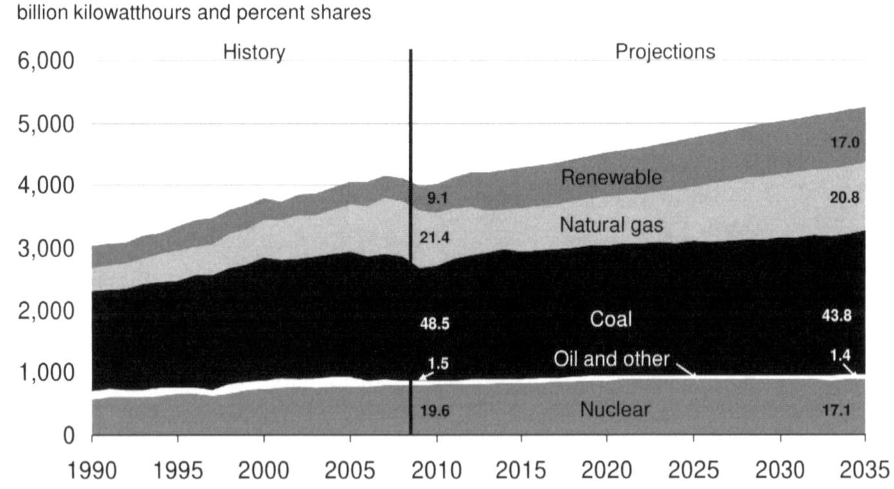

Source: *Annual Energy Outlook 2010*

Figure 2. Major Sources of CO_2 Emissions – 2008 & 2035

Growth in energy-related CO_2 is driven by electricity and transportation fuel use

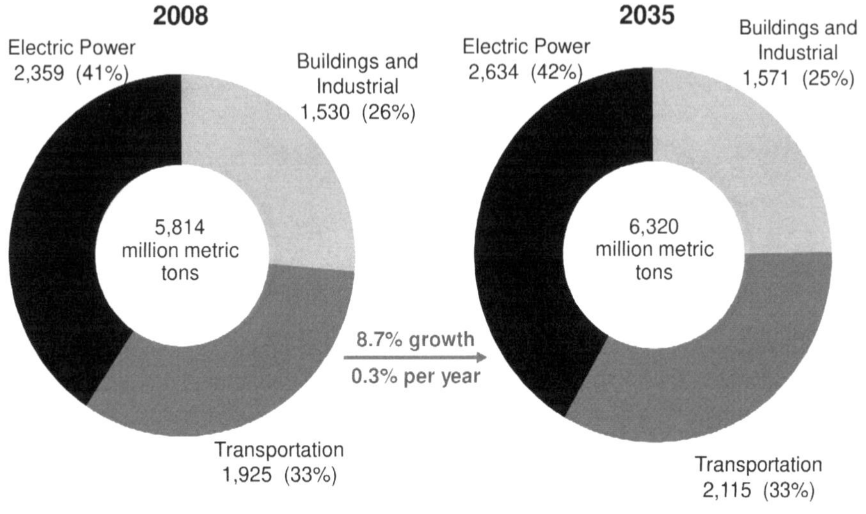

Source: *Annual Energy Outlook 2010*

Value of Carbon

The basic question that a company faces today in being able to move forward aggressively in developing CCS as a viable business is: What will be the value that our society places on carbon to reduce emissions, whether regulatory or voluntary? Without a substantial increase in the economic value of carbon reductions, no significant advances will be made in CCS. Historically, since the passage of the Kyoto accord, carbon values have been relatively volatile. Figure 3 shows the price volatility for both European Allowances (EUAs) and Certified Emission Reduction Credits (CERs) for 2009. With the recent outcome of the Copenhagen meeting, carbon prices in Europe have declined and at the end of 2009, December 2010 EUAs were trading at €12.53/metric ton or $17.82/metric ton on the European Climate Exchange in London. The UN's CERs for delivery in 2010 fell 7.2 percent after the close of the December 2009 Copenhagen meeting and brought prices down to €10.98/metric ton or $15.62/metric ton.

Figure 3. Price Volatility of EUAs & CERs

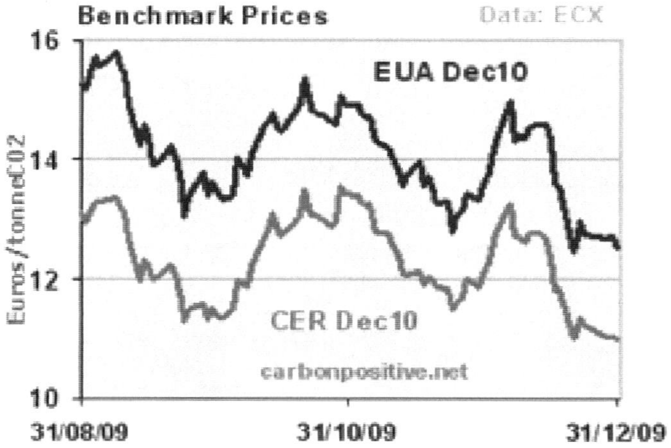

In the U.S. we have one regulated market, that being the Regional Greenhouse Gas Initiative (RGGI) covering 10 northeastern states. These states formed RGGI to achieve a 10% reduction of CO2 emissions by 2018 from the power plants operating within the region. The value of RGGI emissions allowances has dropped significantly by about 50% since they first started trading in September 2008. The first RGGI auction for emission allowances was held on September 25, 2008 and the clearing price for the first allowances sold was $3.07 with 12,565,387 being sold. On December 17, 2009, RGGI 2009 allowances traded at $2.13 and RGGI 2010 allowances traded at $2.07 on the Chicago Climate Futures Exchange. This volatility is primarily due to the uncertainty over whether RGGI emission allowances will have value in an eventual US cap and trade scheme. Table 1 provides the value of the RGGI emission allowances as compared to the global carbon market as reported by Point Carbon.

Table 1. Comparison of Global and US Carbon Markets

	Year	Million m-tons	Value - $ millions	Ave Price $/m-ton
Global (Point Carbon)	2007	2,667	$58,000	$21.75
	2008	4,881	$133,000	$27.25
	2009	8,200	$136,000	$16.58
US-RGGI (RGGI)	2008	44	$145	$2.98
	2009	127	$349	$2.48

The U.S. also has an active voluntary market for verified emission reductions (VERs) and these are traded primarily in the OTC market. The VER-OTC market has experienced similar price erosion but not to the extent that the RGGI allowances have. It its thought that the VER market has experienced only minor price erosion due to the transparency that surrounds these emission reductions and the scrutiny they undergo to be placed on independent Registries.

Cost of Carbon Capture & Sequestration (CCS)

The review above provides us a basis of the value that the European and U.S. societies have placed on emission reductions. Now we need to understand what the cost is to capture CO2 from industrial processes to generate the desired emissions reductions. Since electric power from coal accounts for 48.5% of the US power consumption and 41% of CO2 emissions, this review will focus of the cost of CO2 capture from coal fired power plants. Several studies have been conducted over the past few years by the Canadian government, U.S. National Energy Technology Laboratory (NETL) and Massachusetts Institute of Technology (MIT). Table 2 presents the results of a 2007 NETL study that looked at the cost of capture from three types of fossil fuel power plants, integrated gasification combined cycle (IGCC), pulverized coal super critical (PCSC) and natural gas combined cycle (NGCC). The NETL report presents a comprehensive view of the cost to capture CO2 from the U.S. existing fleet of fossil fueled power plants, PCSC and NGCC, and an IGCC unit from GE Energy (GEE). The results shown are based on new build, state of the art plants. One can see that even with CO2 capture, the NGCC plant has the highest efficiency at 43.7 percent and is the least capital intensive of the capture technologies. Even though the NGCC has the lowest capital it has the highest cost of carbon captured, $70/ton, mainly due to the much smaller amount of CO2 generated by NGCC as compared to PCSC at $45/ton and the GEE IGCC at $27/ton.The Canadian study looked at several configurations of pulverized coal supercritical fired power plants fueled with sub-bituminous coal, bituminous coal and lignite. For this review only two of the plant configurations that were studied are provided, those being for a new build PCSC equipped with amine CO2 capture and a PCSC plant that has been retrofitted with amine CO2 capture. These two plant configurations represent the majority of existing PCSC plants that are operating or could possibly be built in the future. Table 3 presents the results of the Canadian study for these two plant configurations.

Table 2. Cost of CO2 Capture from Fossil Fuel Energy Plants[1]

Fossil Fuel Power Plants	GE Energy IGCC		Pulverized Coal Supercritical		Natural Gas Combined Cycle	
CO2 Capture	No	Yes 90%	No	Yes 90%	No	Yes 90%
Gross Power Output (MW)	770.3	745.0	580.3	663.4	570.2	520.1
Aux. Power Requirement (MW)	130.1	189.2	30.1	117.5	9.8	38.2
Net Power Output (MW)	640.2	555.7	550.2	545.9	560.4	481.9
Net Plant HHV Eff. (%)	38.2	32.5	39.1	27.2	50.8	43.7
Total Plant Cost ($ Billions)	1.161	1.328	0.866	1.567	0.311	0.565
Total Plant Cost ($/kW)	1,813	2,390	1,575	2,870	554	1,172
LCOE (mills/kWh)[a]	78.0	102.9	63.3	114.8	68.4	97.4
Increase in COE (%)	-	32	-	82	-	42
CO2 Emissions (Million tons/yr) @CF[B]	3.937	0.401	3.631	0.516	1.662	0.166
CO2 Captured (Million tons/yr)	-	3.609	-	5.094	-	1.496
Cost of CO2 Captured ($/ ton)[a]	-	27	-	45	-	70

A. Based on coal (Illinois No. 6) at $1.80/MMBtu and natural gas at $6.75/MMBtu and includes cost of compression, transportation for 50 miles and storage in geologic formation with 30 years of monitoring.

B. Capacity (CF) is 80% for IGCC cases and 85% for PCSC and NGCC cases.

Table 3. Cost of CO2 Capture from New Built and Retrofitted PCSC Power Plants[2]

Fossil Fuel Power Plants	Pulverized Coal Supercritical New Build		Pulverized Coal Supercritical Retrofit
CO2 Capture	No	Yes 90%	Yes 90%
Gross Power Output (MW)	542.0	480.5	484.1
Aux. Power Requirement (MW)	38.6	89.2	90.0
Net Power Output (MW)	503.4	391.3	394.1
Net Plant HHV Eff. (%)	42.9	33.4	33.6
Total Plant Cost (CA$ Billions)[c]	1.758	2.320	2.331
Total Plant Cost (CA$/kW)	3,492	5,929	5,915
LCOE (mills/kWh)[d]	64.8	98.3	99.2
Increase in COE (%)	-	52	53
CO2 Emissions (Million metric ton/yr)	2.961	0.388	0.388
CO2 Captured (Million metric tons/yr)	-	2.573	2.573
Cost of CO2 Captured (CA$/metric ton)[e]	-	48.82	50.04

C. Exchange Rate: $1.00 US = $1.15 CA; D. Based on CA$14/tonne sub-bituminous coal (18.7 MJ/kg)

E. Cost includes CO2 capture and compression; no pipeline costs are included

Using the two studies reviewed above one can begin to assign a cost for capturing CO2 from PCSC power plants and IGCC plants. For CO2 capture from a PCSC plant the two reports reached very similar conclusions as to the cost of capture. The NETL cost for capture compression and transport was US$45/ton while the Canadian study showed a cost of CA$48.82/ metric ton for capture and compression only. When the Canadian costs are converted into US$/ton it results in a cost of US$38.20/ton. To put these two studies on the same basis the Canadian study needs to have the cost of transportation and storage added.

In the past couple of years several studies have been conducted to determine the cost of building and operating CO2 pipelines. One such study[3] was issued by the Congressional Research Service (CRS) in January 2008 in which it reviewed what the possible scenarios for CO2 pipeline development should be considered for the future build out of a CO2 pipeline system to service the majority of the pulverized coal power plants in the US. The study looked at coal fired power plants since they contribute approximately one third of the U.S. emissions from fossil fuels. The CRS concluded that there is a considerable amount of uncertainty as to the length of CO2 pipelines that will be required to service coal fired power plants. As a result they relied on a 2005 study which concluded that 77% of the total annual CO2 captured from major North American sources could be stored in reservoirs directly underlying the sources and that an additional 11% could be stored within 100 miles of the original sources.[4] The Midwest Regional Carbon Sequestration Partnership (MRCSP) also studied the issue of CO2 transportation because a large number of coal fired power plants reside in the Partnership's region as do a varied mix of potential geologic sequestrations sites. The MSCSP concluded that CO2 from the power plants could be transported to sequestration sites within 32 miles for most of the power plants.

Based on these two studies, let us assume that CO2 from an emission source will have to be transported 50 miles to a sequestration site. The Canadian study was based on PCSC power plants having a gross power output of 480 MW while the NETL study was based on a PCSC plant with a gross power output of 580 to 663 MW which more accurately reflect the typical size of PCSC plants in operation in the US and results in having larger quantities of captured CO2. For the purpose of this analysis we will use the larger quantity of CO2 captured, 5 million tons per year, to determine the size of CO2 pipeline. Five million tons per year of CO2 results in a daily capture of 266 MMSCF assuming that the plant has an operating rate of 90%. A 16 inch diameter pipeline will be required to transport this volume of CO2. The CRS study relied on cost models developed at Carnegie Mellon University (CMU) to determine the capital cost to construct and operate the CO2 pipelines.[5] The CMU study is based on 2004 cost data for natural gas pipelines and looked at four cost components, right of way (ROW), materials, engineering and overheads and labor, to build a model to estimate the capital cost to construct CO2 pipelines. Figure 3 shows these various cost components for a 16 inch diameter pipeline.

Figure 3. Capital Cost of a 16 inch CO2 Pipeline of Various Lengths in the Midwest[5]

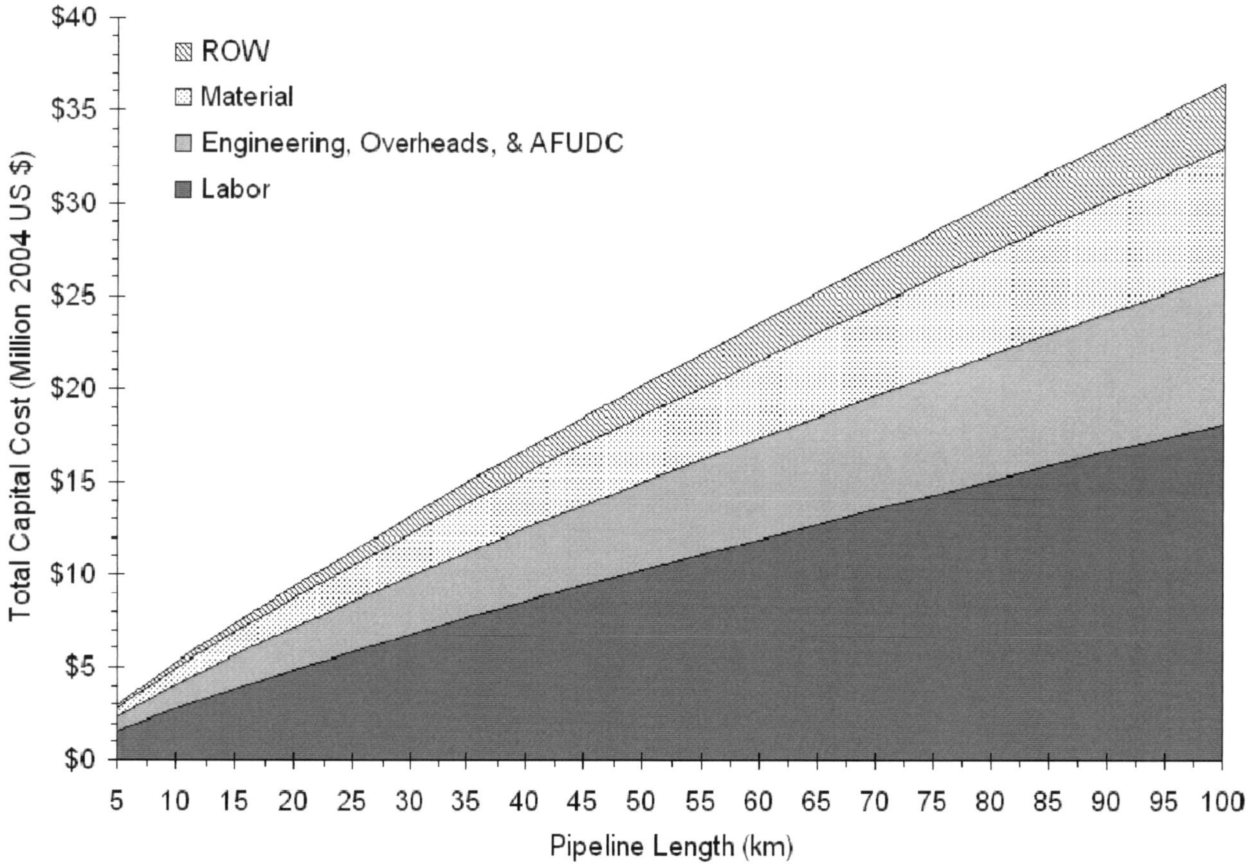

More recent cost data based on actual pipeline cost studies completed in the fall of 2009 indicate that the total cost to construct a 50 mile, 16 inch diameter pipeline would be $49,600,000. This is about three times the cost indicated in the model developed by CMU and shown in Figure 3 above. This cost difference is directly attributable to the inflation in commodities, energy and labor that the US and the rest of the world have experienced since 2007. Prior to the occurrence of this recent inflationary period, the CMU model tracked very well with data from as built pipelines. If we assume a 30 year life of the pipeline, then the cost of the line on a per ton basis will be approximately $0.55 per ton.

The US presently has a network of about 3,500 miles of CO2 pipelines which are primarily concentrated in West Texas and the Rocky Mountain west. The single largest network of pipelines has evolved to take CO2 from natural CO2 reservoirs in southwestern Colorado, southeastern Colorado and northeastern New Mexico to the oil fields of West Texas, with over 1.2 BCF currently being used for EOR in the Permian Basin. Other major areas of pipeline development have occurred in the Gulf Coast region from Mississippi to Louisiana with eventual expansion into Texas near Houston. A significant pipeline system has evolved in Wyoming to take CO2 from ExxonMobil's Shute Creek natural gas processing plant to numerous fields in the Great Divide, Powder River and Wind River Basins. Other smaller regional lines have evolved to transport anthropogenic CO2 from a coal gasification plant (DGC in North Dakota) and ammonia plants (Agrium in Texas and Koch in Oklahoma). Figure 4 provides an overview of the existing CO2 pipeline network. Most of these pipelines have been in operation for over 25 years and have a different cost basis as compared to new lines that will have to be built to transport CO2 from "new" anthropogenic sources of CO2.

Figure 4. Existing US CO2 Pipeline System

In August 2003, the Massachusetts Institute of Technology published the report, "The Economics of CO2 Storage" that investigated the technologies required and associated costs to store CO2 in various geologic strata including depleted oil and gas reservoirs and saline aquifers. Tables 4 below present the results for three of the cases studied.

Table 4. Storage Costs for Depleted Gas Reservoirs: Base, High and Low Cost Cases[6]

Parameter	Units	Base Case	High Cost Case	Low Cost Case
Depleted Gas Reservoir	$/tonne	$4.87	$19.43	$1.20
Depleted Oil Reservoir	$/tonne	$3.82	$11.16	$1.21
Saline Aquifer	$/tonne	$2.93	$11.71	$1.14

More recently in June of 2008 the US Environmental Protection Agency (EPA) published the Technical Support Document, "Geologic CO2 Sequestration Technology and Cost Analysis" that analyzed the various technologies and costs required to sequester and monitor CO2 in geologic formations and depleted oil and gas reservoirs. This report provides design parameters and typical costs for each so that one can design a geological storage site and develop capital costs and operating expenses for a specific sequestration site.

Table 5 combines the pipeline cost derived above and the storage costs from the MIT study and the Canadian CO2 capture study for a direct comparison between it and the NETL study.

Table 5. Comparison of Costs for CCS for A New PCSC Power Plant: $/ton Basis[F]

Study	CO2 Capture & Compression	Pipeline	Storage	Total[F]
NETL	$45.00	Incl.	Incl.	$45.00
Canada	$42.45	$0.55	$4.38	$47.38

F Estimates are based References: 1, 2, 3, 4, 5 & 6.

From Table 5 we have a good estimate of the total cost to capture, compress, transport and store CO2 for long term geological sequestration. These costs are based on studies done from 2003 through 2007 and should be considered as initial estimates but for this analysis provide us with a good approximation of the costs that would be incurred. They do not include any profit that a company entering into the business would expect to obtain to compensate it for the capital employed and provide a nominal rate of return. With the price escalation that industry experienced in 2008 and 2009, these costs are probably 20% to 25% lower than if they would be done today.

Revenue Sources for CCS

In the US at the present time there are two potential revenue sources that can be generated from CCS. First and foremost is from the commodity sale of anthropogenic CO2 for use in enhanced oil recovery (EOR). Presently in the US about 600 MMSCF/D of anthropogenic CO2 is captured from gas processing plants, coal gasification plant and ammonia plants and sold for EOR for a price in the range of $1.25/MCF or $21.85/ton. Of this 600 MMSCF/D of anthropogenic CO2 only about half is monitored under strict protocols and is certified by third party engineering companies that the CO2 has been sequestered. Second is the revenue that results through the sale of the verified emission reduction (VER) credits into the voluntary over the counter (OTC) carbon markets. These OTC carbon credits have a higher market value than the RGGI allowances or credits sold on the Chicago Climate Exchange (CCX). Values for these CCS VER credits in the OTC carbon market ranged from $4 to $6 per metric ton in 2009 which was slightly lower than values achieved in 2008. The lower values obtained in 2009 can be attributed to the uncertainty surrounding the potential regulation of CO2 emissions. When these two revenue sources for CO2 are combined, a producer of anthropogenic has the potential of obtaining from $25.85/ton to $27.85/ton of CO2 sequestered. This represents only about 60% of the cost to capture, compress, transport and sequester CO2 as shown in Table 5.

It is apparent that if CCS is to occur on a broad basis for major anthropogenic sources either additional revenues are required to cover the costs or the additional costs will have to be passed on to consumers in the form of higher energy costs. Several bills were submitted by Congress in 2009 which addressed the need for CCS and offered various solutions to move CCS forward on a broad scale.

1. H.R. 2454, "American Clean Energy and Security Act of 2009" (Waxman Markey bill)
    a. Facilitates carbon capture, transportation and storage
    b. Provides bonus allowances as high as $100/ton for early movers
    c. Recognizes sequestration mediums: saline formations, depleted oil & gas fields and coal seams
    d. Use of EOR as storage results in reduced bonus allowance vales at EPA's discretion

2. SB 1733, "Clean Energy Jobs and American Power Act" (Kerry Boxer bill)
    a. Facilitates carbon capture, transportation and storage
    b. Provides bonus allowances as high as $106/ton for early movers
    c. Recognizes storage as "permanent geological sequestration"
    d. May not recognize EOR as "permanent geological sequestration" since the bill does not apply to Class II wells

3. SB 1013, ""Department of Energy Carbon Capture and Sequestration Program Amendment Act of 2009 (Bingaman Murkowski)
    a. Facilitates carbon capture, transportation and storage
    b. Provides national indemnity program for up to 10 commercial scale CCS projects
    c. Recognizes sequestration mediums: saline formations, depleted oil & gas fields and stacked storage mediums
    d. Sites must comply with Federal and state regulations
    e. May be some financial assistance

4. SB 1502, "Carbon Storage Stewardship Trust Fund Act of 2009", (Casey and Enzi)
    a. Facilitates carbon capture, transportation and storage
    b. Recognizes storage in saline formations hydrocarbon formations, basalt formations, salt caverns, coal seams and any other geologic formation capable of permanent storage
    c. Requires private liability insurance for transport, injection, storage and post closure monitoring
    d. Establishes mechanism to transfer long term liability and stewardship of stored CO2 to the Federal government
    e. Requires fees to be paid to a Federal Trust fund for use by the Federal government upon transfer of the stewardship responsibilities

None of these bills passed and it appears doubtful that any progress will be made in 2010. There are funds available on a limited basis under the U.S. Department of Energy –

National Energy Technology Laboratory Recovery Act: Carbon Capture and Sequestration from Industrial Sources and Innovative Concepts for Beneficial Use of CO2. As part of the Recovery Act, the DOE has $1.3 billion available for supporting new technology and innovation. The awarding of these funds will occur through a two phase program which will ultimately provide up to 50% of the funds required for up to six projects. While this DOE funding is a start, it really does not provide a basis from which broad based commercial development of CCS will occur.

The only assured Federal assistance that exists at the present time for CCS comes from Section 45Q of the Internal Revenue Code (as amended by the Energy Improvement and Extension Act of 2008) that provides for tax credits in relation to CO2 sequestration. Section 45Q allows for a tax credit equal to $10 or $20 per metric ton of qualified CO2 sequestered depending on whether the captured CO2 is used by the taxpayer as a tertiary injectant in a qualified enhanced oil or natural gas recovery project ($10 credit) or disposed of in secure geological storage ($20 credit). The tax credit is limited only to CO2 captured and disposed of or used within the United States. The tax credit can provide an additional source of revenue for CCS assuming that the company has a tax liability.

The tax credit helps to close the gap with respect to the actual costs that will be incurred in the process of CCS but still does not generate enough revenues for a commercial business to be established, especially with the risks and uncertainties that are associated with CCS. There needs to be a way for society to provide additional financial incentives for CCS to proceed besides the current revenue streams available; commodity value for use in EOR, emission reduction credit in the voluntary carbon market and the Section 45Q tax credits. The additional support can come from additional value being assigned to emission reduction credits in the voluntary market, from a pure carbon tax placed on emissions which, if high enough, would compel companies to incur the cost to reduce emissions by CCS, or by cap and trade regulations being implemented that forces emitters to purchase offsets or reduce emissions. Until the gap between value placed on emission reductions and the cost to achieve CCS is overcome, it will be difficult for commercial CCS operations to evolve.

References

1. "Cost and Performance Baseline for Fossil Energy Plants", *Final Report DOE/NETL-2007/1282*
2. "Future CO2 Capture Technology Options for the Canadian Market", *Report No. Coal R309 BERR/Pub URN 07/1251*
3. Paul W. Parfomak and Peter Folger, "Pipelines for Carbon Dioxide (CO2) Control: Network Needs and Cost Uncertainties", *CRS Report for Congress* (January 10, 2008).
4. R.T. Dahowski, J.J. Dooley, C.L. Davidson, S. Bachu, N. Gupta, and J. Gale, "A North American CO2 Storage Supply Curve: Key Findings and Implications for the Cost of CCS Deployment," *Proceedings of the Fifth Annual Conference on Carbon Capture and Sequestration* (Alexandria, VA: May 2-5, 2005).

5. Sean T. McCoy and Edward S. Rubin, "An Engineering-Economic Model of Pipeline Transport of CO2 with Application to Carbon Capture and Storage," *International Journal of Greenhouse Gas Control*, (November 19, 2007).
6. Gemma Heddle, Howard Herzog & Michael Klett, "The Economics of CO2 Storage", *MIT LFEE 2003-003 RP,* August 2003.

AICHE Abstract No. 176624

GHG Emission Reporting Rule

Mary Ellen Ternes

McAfee & Taft

David Winfrey

McAfee & Taft

Kristine Baranski

Cardinal Engineering

Deanne Hughes

Cardinal Engineering

Prepared for Presentation at the 2010 Spring National Meeting

San Antonio, TX, March 22, 2010

GHG Emission Reporting Rule

Mary Ellen Ternes
McAfee & Taft

David Winfrey
McAfee & Taft

Kristine Baranski
Cardinal Engineering

Deanne Hughes
Cardinal Engineering

Abstract: This workshop session created specifically for refiners provides a hands-on review and discussion of EPA's Mandatory Reporting of Greenhouse Gases (GHG) Rule effective December 29, 2009, and specifically, those provisions applicable to refinery facilities. The workshop will be facilitated by expert environmental legal counsel and technical consultants. The format will include a review of the purpose of the rule, GHG Monitoring Plan requirements, transitioning from Best Available Monitoring Methods to the specific GHG monitoring and emission calculation methods required by Subpart Y, quality assurance, missing data calculations, record retention, designated representatives and agreements, and enforcement. The review will also work through compliance with Subpart C (stationary fuel combustion) and Subpart MM (petroleum supplier) requirements. Questions and discussion will be encouraged throughout the review. With this abstract, the Session Chair requests that, prior to March 2010, attendees e-mail specific questions or areas of discussion to be included in the review to Maryellen.ternes@mcafeetaft.com identifying the Subject as: "AICHE Refiners GHG Workshop – Question."

GHG Emission Reporting Rule

Mary Ellen Ternes

McAfee & Taft

David Winfrey

McAfee & Taft

Kristine Baranski

Cardinal Engineering

Deanne Hughes

Cardinal Engineering

Did January 1, 2010 sneak up on anyone? After toasting the New Year, it was time to start monitoring greenhouse gas emissions - or at least the numerous parameters to be used in the calculations of greenhouse gas emissions - for numerous facilities. How is that going for everyone?

The Mandatory Greenhouse Gas Reporting Rule, or Code of Federal Regulations Title 40 Part 98, became effective December 29, 2009. The intent of the rule is to gather greenhouse gas emission data from large emitters and/or generators of greenhouse gases such as carbon dioxide, methane, nitrous oxide, hydrofluorocarbons, perfluorocarbons, sulfur hexafluoride, nitrogen trifluoride, and hydrofluorinated ethers. The data will then be used in future greenhouse gas-related policy. The rule requires particular facilities to report these emissions annually.

The first report, covering emissions from calendar year 2010, is due March 31, 2011 and will cover calendar year 2010. Data will be electronically submitted to the U.S. Environmental Protection Agency (EPA). EPA believes that this rule will apply to approximately 10,000 facilities and address nearly 85% of the nation's greenhouse gas emissions. **Figure 1** below shows the distribution of facilities that EPA anticipates will be covered by the reporting rule.

Figure 1

Number and Distribution of U.S. Facilities Covered by the Mandatory Reporting of Greenhouse Gases Rule

(U.S. Environmental Protection Agency)

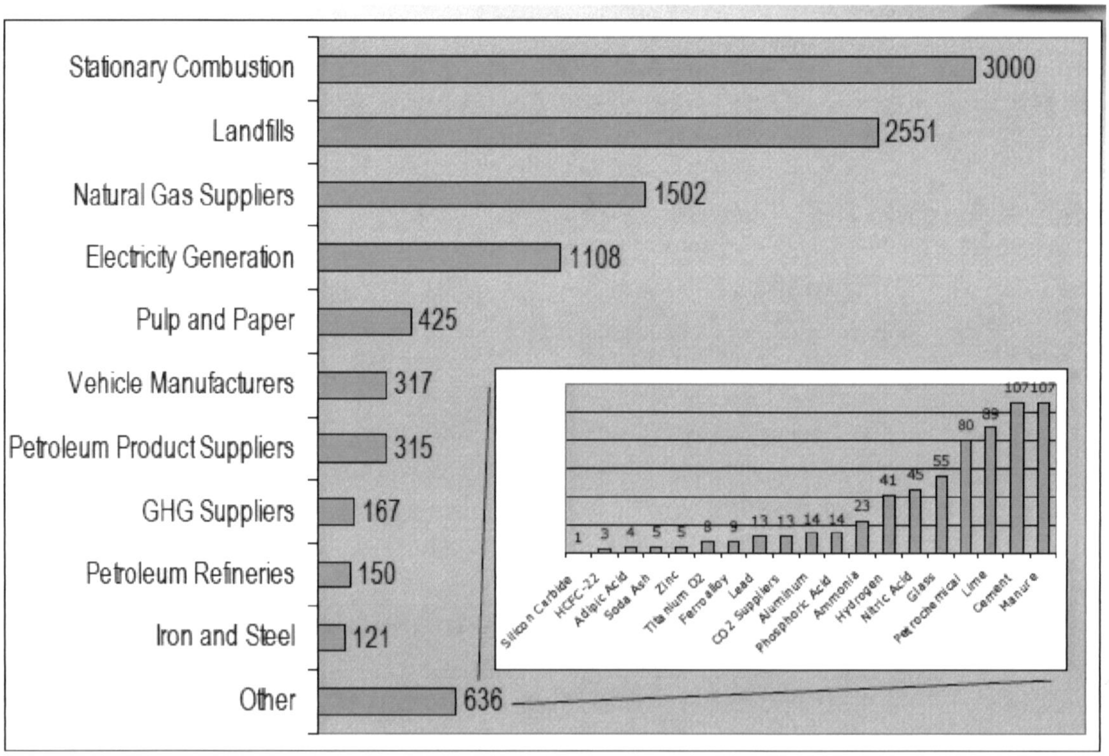

Emissions will be aggregated and reported in units of CO_2e metric tons in addition to metric tons of each individual greenhouse gas. CO_2e is carbon dioxide equivalent, which is an adjusted measurement of total greenhouse gas emissions. It is calculated by adjusting for different global warming potentials (GWP) between the various greenhouse gases. For example, since methane traps 21 times more radiation than carbon dioxide, the current Greenhouse Gas Reporting Rule uses a GWP of 21 tons CO_2e per ton of methane.

This rule does NOT limit greenhouse gas emissions (GHG). For some types of facilities, there is a reporting threshold of 25,000 metric tons per year of CO_2e. However, this threshold is only for use in determining whether these facilities are subject to monitoring and reporting. The threshold is not a limit or "cap" in any way. This rule is also NOT related in any way to any existing state or regional greenhouse gas programs. At this time, the reporting will be completely independent from reporting for any existing state or regional reporting programs. Finally, this rule is NOT a permit of any kind. A separate EPA-proposed Tailoring Rule involves permitting related to greenhouse gases, discussed later in this article, but this is not tied to the reporting rule.

There are three categories of facilities that may be subject: All-in, Threshold, and Supplier categories. The All-in categories, unless further specified, are subject to this rule regardless of the quantity of GHG emissions. Additionally, these facilities must report all emissions for which there is a calculation method in the rule. For example, a petroleum refinery (listed in the "all-in" category) would be required to report not only the emissions as calculated in the petroleum refinery subpart but also any other emissions due to stationary combustion regardless of the quantity of GHG emissions from stationary combustion. For the threshold categories, the facilities that emit more than 25,000 metric tons of CO_2e per year are subject. For comparison to the threshold, all emissions for which there is a calculation method must be aggregated. For suppliers, the facilities must report all emissions according to the appropriate supplier subpart regardless of the quantity of GHG emissions. See Table 1 below for the categories.

Table 1
Greenhouse Gas Reporting Rule Facility Categories

All-In	**Threshold**	**Suppliers**
Electricity Generation	Stationary Combustion Units	Coal to Liquids
Adipic Acid Production	Ferroalloy Production	Petroleum Products
Aluminum Production	Glass Production	Natural Gas and Natural Gas Liquids
Ammonia Manufacturing	Hydrogen Production	Industrial GHG
Cement Production	Iron and Steel Production	Carbon Dioxide
HCFC-22 Production	Lead Production	
HFC-23 Destruction Processes	Pulp and Paper Manufacturing	
Lime Manufacturing	Zinc Production	
Nitric Acid Production		
Petrochemical Production		
Petroleum Refineries		
Phosphoric Acid Production		
Silicon Carbide Production		
Soda Ash Production		
Titanium Dioxide Production		
Municipal Solid Waste Landfills that generate CH_4 equivalent to 25,000 metric tons CO_2 e or more per year		

Manure Management Systems		

There is actually a fourth category – "not yet subject." The originally proposed reporting rule included subparts for twelve additional facility/emission types that were not included in the final rule that became effective on December 29, 2009. In order to adequately address these facilities, additional evaluation was deemed necessary. The subparts that have not yet been finalized are as follows:

•Electronics manufacturing

•Ethanol production

•Fluorinated GHG production

•Food processing

•Magnesium production

•Oil and natural gas systems

•Sulfur hexafluoride (SF_6) from electrical equipment

•Underground coal mines

•Industrial landfills

•Wastewater treatment

•Suppliers of coal

•Geologic sequestration

It is anticipated that the Oil and Natural Gas Systems as well as the Geologic Sequestration (which includes enhanced oil recovery) will be proposed again early in 2010.

Ideally each facility has already determined if the facility/supplier is subject to the reporting rule. If subject, by now the facility should:

1. Have a Monitoring Plan in place.
2. Be tracking data according to the Monitoring Plan.
3. If using Best Available Monitoring Method (BAMM), understand clearly when this is no longer acceptable.
4. Begin preparing for reporting.

Monitoring Plan

The Monitoring Plan is a living document that describes the calculation methods that will be used for reporting GHG emissions, the data required for those calculations, the data gathering equipment calibration / maintenance / repair procedures, and the party who will be responsible for gathering each data component. The Monitoring Plan may refer to other existing documents where specific details are already described. In many cases, facilities will already be monitoring most or all of the data that is required. The responsible party may be either an individual's name or job title and may be different for various parameters as desired.

EPA has indicated that this document should be in place prior to April 1, 2010. Although it is not necessary to submit this plan to EPA, in the event of an audit this document will likely need to be produced. Additionally, any data or documentation that is described or included in the Plan would also potentially need to be produced in the event of an audit.

For facilities that are not subject to reporting but have emissions approaching the applicable threshold, it may be prudent to consider developing a monitoring plan. If a facility determines that they are not subject under a current operating scenario but during the year begins operating such that a threshold will be exceeded by the end of the reporting year, then the facility will be responsible for reporting at the point in time when the operational change occurred.

Data Tracking

Special cases exist for monitoring and reporting in 2010. The Best Available Monitoring Method option is discussed in more detail below. For stationary combustion, an abbreviated emissions report may be submitted for the monitoring year 2010. This report allows for reporting aggregated facility emissions instead of unit level emissions. Also for stationary combustion, there are common stack and common pipe reporting options. The common stack method allows for aggregating multiple combustion units if they share a common duct/stack with a CEMS installed on that duct/stack. The common pipe option allows for aggregating multiple combustion units if they share a fuel supply pipe that is monitored by a fuel flow meter.

For those of you that have all of this figured out already and are tracking your data, here are some tips and suggestions:

- Set up mock calculations or previous years' calculations to ensure that you really are collecting the required data. The best way to ensure that directions are complete is to try to actually do it.
- Remember to update your Monitoring Plan to reflect any changes as needed.
- For facilities required to report emissions due to Stationary Combustion, ensure that all combustion sources are included (space heaters included) unless specifically exempt. There is no de minimis threshold.

- For stationary combustion units that combust multiple fuel types, the calculation Tier is specific to the fuel type. That is, a single unit may use one Tier calculation methodology for one fuel type and another Tier for a second fuel type. Alternatively, a facility may elect to monitor and report using the same Tier for all fuel types combusted provided that the highest applicable Tier is used.
- Consider options to allow you to simplify monitoring – is your plan as streamlined as possible? Are there opportunities to utilize the common pipe or common stack alternative methodologies as allowed for Stationary Combustion? Are there opportunities to utilize existing plant data rather than tracking additional parameters?
- Are any changes on the horizon for your facility that might impact your status (subject or not subject, subpart applicability, etc.)?
- Be prepared for additional subparts that may impact your facility. EPA has indicated that the Oil and Natural Gas as well as the Geological Sequestration subparts may be proposed in early 2010.
- Continue to monitor the EPA website for updates and clarifications. As EPA receives multiple questions on a similar topic, these are posted to the "FAQ" section. Similar facilities likely have similar questions and concerns as Monitoring Plans are developed.
- As discussed previously, much of the data required is likely already being tracked elsewhere already. Take advantage of existing data tracking methods. Some additional calculations may be required, but the data tracking may not require extensive changes.
- Ensure that all calibration records, as well as monitoring equipment maintenance and repair records are retained for the minimum time specified (generally three years).
- If there are any periods of missing data, then the appropriate missing data procedures (as specified in the reporting rule) must be utilized and documented.

BAMM/Abbreviated Report Discussion

From January 1, 2010 through March 31, 2010, Best Available Monitoring Methods may be used in lieu of the required methods. This would apply to facilities that require additional equipment to monitor all of the required parameters, facilities that require different monitoring equipment to achieve the required precision/accuracy, and facilities that have not performed the required calibrations. Unless an extension request has been submitted on or before January 28, 2010 and approved, the specified methods must be utilized by April 1, 2010. If an extension request has been submitted and approved, the extension will have a deadline no later than December 31, 2010. In some cases, BAMM extensions may not even be required. For example, EPA has clarified that in the event that initial calibrations are not complete for continuously operating unit/meter with infrequent outages, the initial calibration may be postponed until the next scheduled maintenance event. This does not require a BAMM extension. Please visit the EPA website for additional details:

http://www.epa.gov/climatechange/emissions/ghg_faq.html

Additionally, for facilities that are ONLY subject to reporting due to Stationary Combustion emissions, an abbreviated emission report may be submitted for the calendar year 2010 emissions. Essentially, the facility may choose to utilize any of the four Tier methodologies that are presented in Subpart C. This is only for calendar year 2010 emissions, and beginning January 1, 2011 the facility would be required to monitor and report based on the appropriate Tier as specified in Subpart C.

Report Requirements and Software

In addition to general requirements that all subject facilities must report, each subpart may have additional requirements. In some cases, these additional requirements are not required to complete the GHG emission calculations but may require information collection throughout the year. EPA is expecting to have the facility and user registration database ready for input by summer 2010. The emissions input system is expected to be operational in January 2011 with training set to begin in the fall of 2010. The certificate of registration (to designate and officially certify the designated representative) must be completed and submitted at least 60 days prior to the initial emission report deadline of March 31, 2011.

In most cases, the reporting will be at the facility level. For some cases, including many of the supplier facility types, reporting will be at the facility and corporate level.

Designated Representative

The reporting rule requires that the reports due March 31, 2011 be submitted by the "Designated Representative." Do not confuse the reporting rule's designated representative with other reporting programs. The designated representative is not the same as the responsible corporate official which signs Title V Certifications of Compliance. The responsible corporate official is just that, generally an employee of a corporation. The designated representative may most commonly be a consultant, and not a corporate employee.

Each reporting facility and reporting supplier must designate a single designated representative by submitting a "certificate of representation" with EPA. This designated representative is responsible for certifying, signing and submitting the GHG emissions reports and any other submissions for such facility and supplier. For reporters subject to 40 C.F.R. Part 75, the same individual signing the Part 75 reports will be the designated representative for purposes of the Reporting Rule.

In order to secure the reporters designated representative, and before filing a certificate of representation, the owners and operators of the reporting facility or supplier must enter into a binding agreement with the designated representative. The purpose of the binding agreement is so that there is a contractual mechanism ensuring the reliability of the reported information and to support potential enforcement. The designated representative must act in compliance with the certification statement requirements, and the acts of the designated representative subject the owners and operators to legal liability for the representations, actions, inactions or submissions of the designated representative. Moreover, an enforcement action or judicial order against the designated representative is also binding upon the designated representative's owners and operators.

Again, the owner and operator must identify the designated representative at least 60 days before the initial reporting deadline, i.e., for the deadline of March 31, 2011, the designated representative must be identified by January 31, 2011. Of course, the designated representative can be changed at anytime with the submittal and effective upon EPA's receipt of a new certificate of representation. And new owners or operators must be included in an updated certificate of representation within 90 days of the change, though, failing to submit updated information does not prevent EPA from binding new owners and operators to the designate representatives representations.

The designated representative can have one alternate representative. This alternate representative can act on behalf of the original designated representative by filing a "certificate of representation" with EPA identifying the alternate. As long as the alternate acts pursuant to applicable procedures, the alternate binds and represents the owners and operators just like the designated representative.

EPA does not wish owners and operators to submit their binding agreements with their designated representative. Moreover, EPA does not care how the owners and operators contractually or otherwise resolve issues regarding selecting the designated representative or any other approach used by the owners and operators to resolve issues amongst themselves regarding compliance with this rule. The rule applies equally to owners and operators, and this means that the owners and operators must resolve these issues on their own.

However, EPA has set specific requirements for the certificate of representation. The certificate of representation must: identify the facility or supplier; identify the owner(s), with controlling interest at a minimum; identify the operator(s); provide the designated representative's specific contact information, as well as contact information for any alternate designated representative. The certificate must also include a certification statement, signed and dated by the designated/alternate representative, as follows:

> I certify that I was selected as the designated representative or alternate designated representative, as applicable, by an agreement binding on the owners and operators of the facility or supplier, as applicable.
>
> I certify that I have all the necessary authority to carry out my duties and responsibilities under 40 CFR part 98 on behalf of the owners and operators of the facility or supplier, as applicable, and that each such owner and operator shall be fully bound by my representations, actions, inactions, or submissions.
>
> I certify that the owners and operators of the facility or supplier, as applicable, shall be bound by any order issued to me by the Administrator or a court regarding the facility or supplier.

> If there are multiple owners and operators of the facility or supplier, as applicable, I certify that I have given a written notice of my selection as the "designated representative" or "alternate designated representative", as applicable, and of the agreement by which I was selected to each owner and operator of the facility or supplier.

Even while EPA has set rigorous requirements for binding agreements and certificates of representation for the designated/alternate representatives, there is some flexibility in the scope of individuals allowed to complete and electronic submission to EPA. Individuals may submit an electronic submission to EPA on behalf of the designated or alternate representative with the electronic submission of a "notice of delegation."

The individual annual GHG emissions report will be submitted electronically using EPA's electronic reporting system currently under development. Each report must be accompanied with the following certification, signed by the designated representative:

> I am authorized to make this submission on behalf of the owners and operators of the facility or supplier, as applicable, for which the submission is made. I certify under penalty of law that I have personally examined, and am familiar with, the statements and information submitted in this document and all its attachments. Based on my inquiry of those individuals with primary responsibility for obtaining the information, I certify that the statements and information are to the best of my knowledge and belief true, accurate, and complete. I am aware that there are significant penalties for submitting false statements and information or omitting required statements and information, including the possibility of fine or imprisonment.

It is easy to understand a reporter wanting to submit its certificate of representation for its designated representative as soon as possible. But we must wait. EPA does not expect to have its electronic reporting system operational until January 2011. However, EPA does intend to make training on the emissions reporting system available as early as the Fall of 2010, and the separate module for registering users and facilities should be available by the Summer of 2010.

Emission Data Verification

In its final rule, EPA decided against third party verification and self verification, finding initial reporter self-certification followed by EPA's verification through data analysis and on-site auditing to be the best choice. Upon receipt of the data, EPA's data analysis may identify errors. EPA will notify the reporter, and the reporter has 45 days to submit a corrected report. The corrected report must correct all errors identified by both EPA and the reporter. Needless to say, sources tempted to bias their emission rates on the high side will face increased scrutiny through the data analysis and on-site auditing, and thus does so at their peril.

Enforcement

EPA will enforce violations of this rule as violations of the Clean Air Act. Violations may include failure to report GHG emissions despite applicability of the rule, failure to collect all data required to calculate GHG emissions pursuant to the applicable subparts, failure to continuously monitor and test as required by the applicable subpart, failure to retain all records required to properly verify the amount of GHG emissions, and finally, failure to calculate GHG emissions as required by the applicable subparts. Each day of violation constitutes a separate violations.

Exiting the Program

While the proposed rule provided a "once-in, always-in" requirement, in the final rule EPA allows facilities and suppliers to escape reporting as long as they report less than 25,000 metric tons of CO_2e for five consecutive years, or less than 15,000 metric tons for 3 consecutive years. Entities can also stop reporting if all GHG emitting sources cease to operate.

Public Availability of Data

EPA will release and publish the emissions reported pursuant to this rule. EPA does not consider emissions data to be "confidential business information" or CBI. Data other than emissions data may very well meet the definition of CBI. EPA has stated its intent to propose data elements which it considers to fall within the scope of "emissions data" that would not be protected by as CBI.

Implications of publicly available emissions data in light of EPA rulemaking, legislation and litigation

GHG emissions data will be publicly released to a troubling landscape for reporting GHG emission sources. First, while COP15 (the December 2009 climate change meeting in Copenhagen) did not directly affect U.S. GHG emission sources, we all have watched and waited on the status of U.S. GHG cap and trade legislation, which is expected to be reviewed again by the Senate in 2010.

There is also pending GHG regulation. On December 15, 2009, EPA issued its final "endangerment finding," concluding that GHG emissions do indeed endanger public health and the environment. As a result, EPA is prepared to issue regulations actually controlling GHG emissions for the first time. Because these new rules will actually control GHG emissions, rather than simply require GHG emissions to be monitored, these new rules will automatically trigger application of the CAA prevention of significant deterioration (PSD) and Title V operating permit programs to major stationary sources of GHG emissions.

The PSD rules include the expensive demonstration, "Best Available Control Technology," or BACT demonstration. The triggering thresholds for PSD and Title V are 100 and 250 tons per year, which is quite low when considering emissions of GHGs. Thus, in September 2009, EPA issued a proposed "tailoring" rule explaining the legal mechanism upon which it would rely to raise the "major stationary source" air pollutant emission threshold of 250 tons per year and 100 tons per year to 25,000 tons per year. With its proposed "tailoring rule," EPA also explained the manner in which it would implement the CAA permitting programs to major stationary source GHG emission sources. Currently, EPA is expected to promulgate its final rules regulating and controlling GHG emissions from light duty vehicles, as well as its final "tailoring rule," in March 2010. Though subject to legal challenge, EPA's rules promulgated at the time of this writing are currently final and effective, and will remain so unless successfully challenged directly in court, or unless Congress adopts legislation preempting EPA's regulatory authority to address GHGs under the Clean Air Act.

In addition to the legislation and regulation, there is also litigation. The case that hits most closely to home most AICHE members is probably *Comer v. Murphy Oil*, where the Fifth Circuit Court of Appeals held that plaintiffs can go to trial against fossil fuel producers to prove that the GHG produced from their production and ultimate use of fossil fuels resulted in GHG emissions which increased the strength of Hurrican Katrina, and as a result, exacerbated property damage to their property from Hurricane Katrina. The plaintiffs are seeking money damages as compensation.

Beyond enforcement scrutiny from compliance with the GHG reporting rule itself, the GHG reporting rule will cause the public release of GHG emissions data at a time when this data could be used to conduct PSD and BACT review, as well as increase scrutiny by GHG plaintiffs' attorneys.

Perhaps there is safety in numbers.

Bibliography

Mandatory Reporting of Greenhouse Gases, 74 Fed. Reg. 56,260 (Oct. 30, 2009).

Comer v. Murphy Oil USA, 2009 U.S. App. LEXIS 22774 (5th Cir. Oct. 16, 2009).

U.S.EPA Special Provisions for 2010 Fact Sheet, available at:
http://www.epa.gov/climatechange/emissions/downloads/2010SpecialProvisionsInfoSheet.pdf

U.S. EPA Frequently Asked Questions, available at:
http://www.epa.gov/climatechange/emissions/ghg_faq.html

U.S. EPA Final Monitoring Rule Overview, available at:
http://www.epa.gov/climatechange/emissions/downloads/FinalMRROverview.pdf

AIChE Paper Number 43a

CO2 Emission Reporting Rule In Ethylene Plants

Mary Ellen Ternes

McAfee & Taft

David Winfrey

McAfee & Taft

Kristine Baranski

Cardinal Engineering

Deanne Hughes

Cardinal Engineering

Prepared for Presentation at the 2010 Spring National Meeting

San Antonio, TX, March 22, 2010

CO2 Emission Reporting Rule In Ethylene Plants

Mary Ellen Ternes

McAfee & Taft

David Winfrey

McAfee & Taft

Kristine Baranski

Cardinal Engineering

Deanne Hughes

Cardinal Engineering

Abstract: This workshop session created specifically for ethylene producers provides a hands-on review and discussion of EPA's Mandatory Reporting of Greenhouse Gases (GHG) Rule effective December 29, 2009, and specifically, those provisions applicable to ethylene production facilities. The workshop will be facilitated by expert environmental legal counsel and technical consultants. As outlined more specifically below, the format will include a review of the purpose of the rule, GHG Monitoring Plan requirements, transitioning from Best Available Monitoring Methods to the specific GHG monitoring and emission calculation methods required by Subpart X, quality assurance, missing data calculations, record retention, designated representatives and agreements, and enforcement. The review will also work through compliance with Subpart C (stationary fuel combustion) and Subpart MM (petroleum supplier) requirements to the extent applicable. Questions and discussion will be encouraged throughout the review. With this abstract, the Session Chair, requests that, prior to March 2010, attendees e-mail specific questions or areas of discussion to be included in the review to Maryellen.ternes@mcafeetaft.com identifying the Subject as: "AICHE Ethylene Producers GHG Workshop – Question."

The petrochemical producer provisions in EPA's GHG Reporting Rule, at 40 CFR Part 98, Subpart X applies to: all processes that produce ethylene, including processes that produce the petrochemical as an intermediate in the onsite production of other chemicals as well as processes that produce the petrochemical as an end product for sale or shipment offsite.

CO2 Emission Reporting Rule In Ethylene Plants

Mary Ellen Ternes

McAfee & Taft

David Winfrey

McAfee & Taft

Kristine Baranski

Cardinal Engineering

Deanne Hughes

Cardinal Engineering

Did January 1, 2010 sneak up on anyone? After toasting the New Year, it was time to start monitoring greenhouse gas emissions - or at least the numerous parameters to be used in the calculations of greenhouse gas emissions - for numerous facilities. How is that going for everyone?

The Mandatory Greenhouse Gas Reporting Rule, or Code of Federal Regulations Title 40 Part 98, became effective December 29, 2009. The intent of the rule is to gather greenhouse gas emission data from large emitters and/or generators of greenhouse gases such as carbon dioxide, methane, nitrous oxide, hydrofluorocarbons, perfluorocarbons, sulfur hexafluoride, nitrogen trifluoride, and hydrofluorinated ethers. The data will then be used in future greenhouse gas-related policy. The rule requires particular facilities to report these emissions annually.

The first report, covering emissions from calendar year 2010, is due March 31, 2011 and will cover calendar year 2010. Data will be electronically submitted to the U.S. Environmental Protection Agency (EPA). EPA believes that this rule will apply to approximately 10,000 facilities and address nearly 85% of the nation's greenhouse gas emissions. **Figure 1** below shows the distribution of facilities that EPA anticipates will be covered by the reporting rule.

Figure 1

Number and Distribution of U.S. Facilities Covered by the Mandatory Reporting of Greenhouse Gases Rule

(U.S. Environmental Protection Agency)

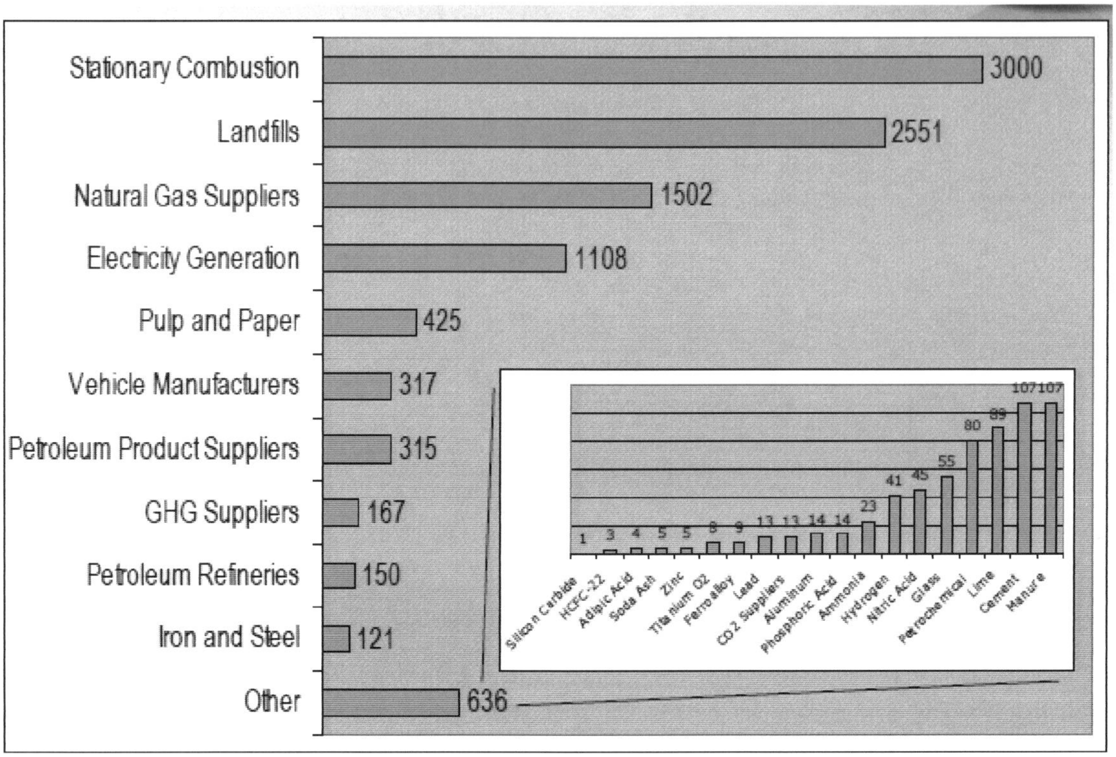

Emissions will be aggregated and reported in units of CO_2e metric tons in addition to metric tons of each individual greenhouse gas. CO_2e is carbon dioxide equivalent, which is an adjusted measurement of total greenhouse gas emissions. It is calculated by adjusting for different global warming potentials (GWP) between the various greenhouse gases. For example, since methane traps 21 times more radiation than carbon dioxide, the current Greenhouse Gas Reporting Rule uses a GWP of 21 tons CO_2e per ton of methane.

This rule does NOT limit greenhouse gas emissions (GHG). For some types of facilities, there is a reporting threshold of 25,000 metric tons per year of CO_2e. However, this threshold is only for use in determining whether these facilities are subject to monitoring and reporting. The threshold is not a limit or "cap" in any way. This rule is also NOT related in any way to any existing state or regional greenhouse gas programs. At this time, the reporting will be completely independent from reporting for any existing state or regional reporting programs. Finally, this rule is NOT a permit of any kind. A separate EPA-proposed Tailoring Rule involves permitting related to greenhouse gases, discussed later in this article, but this is not tied to the reporting rule.

There are three categories of facilities that may be subject: All-in, Threshold, and Supplier categories. The All-in categories, unless further specified, are subject to this rule regardless of the quantity of GHG emissions. Additionally, these facilities must report all emissions for which there is a calculation method in the rule. For example, a petroleum refinery (listed in the "all-in" category) would be required to report not only the emissions as calculated in the petroleum refinery subpart but also any other emissions due to stationary combustion regardless of the quantity of GHG emissions from stationary combustion. For the threshold categories, the facilities that emit more than 25,000 metric tons of CO_2e per year are subject. For comparison to the threshold, all emissions for which there is a calculation method must be aggregated. For suppliers, the facilities must report all emissions according to the appropriate supplier subpart regardless of the quantity of GHG emissions. See Table 1 below for the categories.

Table 1
Greenhouse Gas Reporting Rule Facility Categories

All-In	**Threshold**	**Suppliers**
Electricity Generation	Stationary Combustion Units	Coal to Liquids
Adipic Acid Production	Ferroalloy Production	Petroleum Products
Aluminum Production	Glass Production	Natural Gas and Natural Gas Liquids
Ammonia Manufacturing	Hydrogen Production	Industrial GHG
Cement Production	Iron and Steel Production	Carbon Dioxide
HCFC-22 Production	Lead Production	
HFC-23 Destruction Processes	Pulp and Paper Manufacturing	
Lime Manufacturing	Zinc Production	
Nitric Acid Production		
Petrochemical Production		
Petroleum Refineries		
Phosphoric Acid Production		
Silicon Carbide Production		
Soda Ash Production		
Titanium Dioxide Production		
Municipal Solid Waste Landfills that generate CH4 equivalent to 25,000 metric tons CO2 e or more per year		

Manure Management Systems		

There is actually a fourth category – "not yet subject." The originally proposed reporting rule included subparts for twelve additional facility/emission types that were not included in the final rule that became effective on December 29, 2009. In order to adequately address these facilities, additional evaluation was deemed necessary. The subparts that have not yet been finalized are as follows:

•Electronics manufacturing

•Ethanol production

•Fluorinated GHG production

•Food processing

•Magnesium production

•Oil and natural gas systems

•Sulfur hexafluoride (SF_6) from electrical equipment

•Underground coal mines

•Industrial landfills

•Wastewater treatment

•Suppliers of coal

•Geologic sequestration

It is anticipated that the Oil and Natural Gas Systems as well as the Geologic Sequestration (which includes enhanced oil recovery) will be proposed again early in 2010.

Ideally each facility has already determined if the facility/supplier is subject to the reporting rule. If subject, by now the facility should:

1. Have a Monitoring Plan in place.
2. Be tracking data according to the Monitoring Plan.
3. If using Best Available Monitoring Method (BAMM), understand clearly when this is no longer acceptable.

4. Begin preparing for reporting.

Monitoring Plan

The Monitoring Plan is a living document that describes the calculation methods that will be used for reporting GHG emissions, the data required for those calculations, the data gathering equipment calibration / maintenance / repair procedures, and the party who will be responsible for gathering each data component. The Monitoring Plan may refer to other existing documents where specific details are already described. In many cases, facilities will already be monitoring most or all of the data that is required. The responsible party may be either an individual's name or job title and may be different for various parameters as desired.

EPA has indicated that this document should be in place prior to April 1, 2010. Although it is not necessary to submit this plan to EPA, in the event of an audit this document will likely need to be produced. Additionally, any data or documentation that is described or included in the Plan would also potentially need to be produced in the event of an audit.

For facilities that are not subject to reporting but have emissions approaching the applicable threshold, it may be prudent to consider developing a monitoring plan. If a facility determines that they are not subject under a current operating scenario but during the year begins operating such that a threshold will be exceeded by the end of the reporting year, then the facility will be responsible for reporting at the point in time when the operational change occurred.

Data Tracking

Special cases exist for monitoring and reporting in 2010. The Best Available Monitoring Method option is discussed in more detail below. For stationary combustion, an abbreviated emissions report may be submitted for the monitoring year 2010. This report allows for reporting aggregated facility emissions instead of unit level emissions. Also for stationary combustion, there are common stack and common pipe reporting options. The common stack method allows for aggregating multiple combustion units if they share a common duct/stack with a CEMS installed on that duct/stack. The common pipe option allows for aggregating multiple combustion units if they share a fuel supply pipe that is monitored by a fuel flow meter.

For those of you that have all of this figured out already and are tracking your data, here are some tips and suggestions:

- Set up mock calculations or previous years' calculations to ensure that you really are collecting the required data. The best way to ensure that directions are complete is to try to actually do it.
- Remember to update your Monitoring Plan to reflect any changes as needed.

- For facilities required to report emissions due to Stationary Combustion, ensure that all combustion sources are included (space heaters included) unless specifically exempt. There is no de minimis threshold.

- For stationary combustion units that combust multiple fuel types, the calculation Tier is specific to the fuel type. That is, a single unit may use one Tier calculation methodology for one fuel type and another Tier for a second fuel type. Alternatively, a facility may elect to monitor and report using the same Tier for all fuel types combusted provided that the highest applicable Tier is used.

- Consider options to allow you to simplify monitoring – is your plan as streamlined as possible? Are there opportunities to utilize the common pipe or common stack alternative methodologies as allowed for Stationary Combustion? Are there opportunities to utilize existing plant data rather than tracking additional parameters?

- Are any changes on the horizon for your facility that might impact your status (subject or not subject, subpart applicability, etc.)?

- Be prepared for additional subparts that may impact your facility. EPA has indicated that the Oil and Natural Gas as well as the Geological Sequestration subparts may be proposed in early 2010.

- Continue to monitor the EPA website for updates and clarifications. As EPA receives multiple questions on a similar topic, these are posted to the "FAQ" section. Similar facilities likely have similar questions and concerns as Monitoring Plans are developed.

- As discussed previously, much of the data required is likely already being tracked elsewhere already. Take advantage of existing data tracking methods. Some additional calculations may be required, but the data tracking may not require extensive changes.

- Ensure that all calibration records, as well as monitoring equipment maintenance and repair records are retained for the minimum time specified (generally three years).

- If there are any periods of missing data, then the appropriate missing data procedures (as specified in the reporting rule) must be utilized and documented.

BAMM/Abbreviated Report Discussion

From January 1, 2010 through March 31, 2010, Best Available Monitoring Methods may be used in lieu of the required methods. This would apply to facilities that require additional equipment to monitor all of the required parameters, facilities that require different monitoring equipment to achieve the required precision/accuracy, and facilities that have not performed the required calibrations. Unless an extension request has been submitted on or before January 28, 2010 and approved, the specified methods must be utilized by April 1, 2010. If an extension request has been submitted and approved, the extension will have a deadline no later than December 31, 2010. In some cases, BAMM extensions may not even be required. For example, EPA has clarified that in the event that initial calibrations are not complete for continuously operating unit/meter with infrequent outages, the initial calibration may be postponed until the next scheduled maintenance event. This does not require a BAMM extension. Please visit the EPA website for additional details:

http://www.epa.gov/climatechange/emissions/ghg_faq.html

Additionally, for facilities that are ONLY subject to reporting due to Stationary Combustion emissions, an abbreviated emission report may be submitted for the calendar year 2010 emissions. Essentially, the facility may choose to utilize any of the four Tier methodologies that are presented in Subpart C. This is only for calendar year 2010 emissions, and beginning January 1, 2011 the facility would be required to monitor and report based on the appropriate Tier as specified in Subpart C.

Report Requirements and Software

In addition to general requirements that all subject facilities must report, each subpart may have additional requirements. In some cases, these additional requirements are not required to complete the GHG emission calculations but may require information collection throughout the year. EPA is expecting to have the facility and user registration database ready for input by summer 2010. The emissions input system is expected to be operational in January 2011 with training set to begin in the fall of 2010. The certificate of registration (to designate and officially certify the designated representative) must be completed and submitted at least 60 days prior to the initial emission report deadline of March 31, 2011.

In most cases, the reporting will be at the facility level. For some cases, including many of the supplier facility types, reporting will be at the facility and corporate level.

Designated Representative

The reporting rule requires that the reports due March 31, 2011 be submitted by the "Designated Representative." Do not confuse the reporting rule's designated representative with other reporting programs. The designated representative is not the same as the responsible corporate official which signs Title V Certifications of Compliance. The responsible corporate official is just that, generally an employee of a corporation. The designated representative may most commonly be a consultant, and not a corporate employee.

Each reporting facility and reporting supplier must designate a single designated representative by submitting a "certificate of representation" with EPA. This designated representative is responsible for certifying, signing and submitting the GHG emissions reports and any other submissions for such facility and supplier. For reporters subject to 40 C.F.R. Part 75, the same individual signing the Part 75 reports will be the designated representative for purposes of the Reporting Rule.

In order to secure the reporters designated representative, and before filing a certificate of representation, the owners and operators of the reporting facility or supplier must enter into a binding agreement with the designated representative. The purpose of the binding agreement is so that there is a contractual mechanism ensuring the reliability of the reported information and to support potential enforcement. The designated representative must act in compliance with the certification statement requirements, and the acts of the designated representative subject the owners and operators to legal liability for the representations, actions, inactions or submissions of the designated representative. Moreover, an enforcement action or judicial order against the designated representative is also binding upon the designated representative's owners and operators.

Again, the owner and operator must identify the designated representative at least 60 days before the initial reporting deadline, i.e., for the deadline of March 31, 2011, the designated representative must be identified by January 31, 2011. Of course, the designated representative can be changed at anytime with the submittal and effective upon EPA's receipt of a new certificate of representation. And new owners or operators must be included in an updated certificate of representation within 90 days of the change, though, failing to submit updated information does not prevent EPA from binding new owners and operators to the designate representatives representations.

The designated representative can have one alternate representative. This alternate representative can act on behalf of the original designated representative by filing a "certificate of representation" with EPA identifying the alternate. As long as the alternate acts pursuant to applicable procedures, the alternate binds and represents the owners and operators just like the designated representative.

EPA does not wish owners and operators to submit their binding agreements with their designated representative. Moreover, EPA does not care how the owners and operators contractually or otherwise resolve issues regarding selecting the designated representative or any other approach used by the owners and operators to resolve issues amongst themselves regarding compliance with this rule. The rule applies equally to owners and operators, and this means that the owners and operators must resolve these issues on their own.

However, EPA has set specific requirements for the certificate of representation. The certificate of representation must: identify the facility or supplier; identify the owner(s), with controlling interest at a minimum; identify the operator(s); provide the designated representative's specific contact information, as well as contact information for any alternate designated representative. The certificate must also include a certification statement, signed and dated by the designated/alternate representative, as follows:

> I certify that I was selected as the designated representative or alternate designated representative, as applicable, by an agreement binding on the owners and operators of the facility or supplier, as applicable.
>
> I certify that I have all the necessary authority to carry out my duties and responsibilities under 40 CFR part 98 on behalf of the owners and operators of the facility or supplier, as applicable, and that each such owner and operator shall be fully bound by my representations, actions, inactions, or submissions.

I certify that the owners and operators of the facility or supplier, as applicable, shall be bound by any order issued to me by the Administrator or a court regarding the facility or supplier.

If there are multiple owners and operators of the facility or supplier, as applicable, I certify that I have given a written notice of my selection as the "designated representative" or "alternate designated representative", as applicable, and of the agreement by which I was selected to each owner and operator of the facility or supplier.

Even while EPA has set rigorous requirements for binding agreements and certificates of representation for the designated/alternate representatives, there is some flexibility in the scope of individuals allowed to complete and electronic submission to EPA. Individuals may submit an electronic submission to EPA on behalf of the designated or alternate representative with the electronic submission of a "notice of delegation."

The individual annual GHG emissions report will be submitted electronically using EPA's electronic reporting system currently under development. Each report must be accompanied with the following certification, signed by the designated representative:

I am authorized to make this submission on behalf of the owners and operators of the facility or supplier, as applicable, for which the submission is made. I certify under penalty of law that I have personally examined, and am familiar with, the statements and information submitted in this document and all its attachments. Based on my inquiry of those individuals with primary responsibility for obtaining the information, I certify that the statements and information are to the best of my knowledge and belief true, accurate, and complete. I am aware that there are significant penalties for submitting false statements and information or omitting required statements and information, including the possibility of fine or imprisonment.

It is easy to understand a reporter wanting to submit its certificate of representation for its designated representative as soon as possible. But we must wait. EPA does not expect to have its electronic reporting system operational until January 2011. However, EPA does intend to make training on the emissions reporting system available as early as the Fall of 2010, and the separate module for registering users and facilities should be available by the Summer of 2010.

Emission Data Verification

In its final rule, EPA decided against third party verification and self verification, finding initial reporter self-certification followed by EPA's verification through data analysis and on-site auditing to be the best choice. Upon receipt of the data, EPA's data analysis may identify errors. EPA will notify the reporter, and the reporter has 45 days to submit a corrected report. The corrected report must correct all errors identified by both EPA and the reporter. Needless to say, sources tempted to bias their emission rates on the high side will face increased scrutiny through the data analysis and on-site auditing, and thus does so at their peril.

Enforcement

EPA will enforce violations of this rule as violations of the Clean Air Act. Violations may include failure to report GHG emissions despite applicability of the rule, failure to collect all data required to calculate GHG emissions pursuant to the applicable subparts, failure to continuously monitor and test as required by the applicable subpart, failure to retain all records required to properly verify the amount of GHG emissions, and finally, failure to calculate GHG emissions as required by the applicable subparts. Each day of violation constitutes a separate violations.

Exiting the Program

While the proposed rule provided a "once-in, always-in" requirement, in the final rule EPA allows facilities and suppliers to escape reporting as long as they report less than 25,000 metric tons of CO_2e for five consecutive years, or less than 15,000 metric tons for 3 consecutive years. Entities can also stop reporting if all GHG emitting sources cease to operate.

Public Availability of Data

EPA will release and publish the emissions reported pursuant to this rule. EPA does not consider emissions data to be "confidential business information" or CBI. Data other than emissions data may very well meet the definition of CBI. EPA has stated its intent to propose data elements which it considers to fall within the scope of "emissions data" that would not be protected by as CBI.

Implications of publicly available emissions data in light of EPA rulemaking, legislation and litigation

GHG emissions data will be publicly released to a troubling landscape for reporting GHG emission sources. First, while COP15 (the December 2009 climate change meeting in Copenhagen) did not directly affect U.S. GHG emission sources, we all have watched and waited on the status of U.S. GHG cap and trade legislation, which is expected to be reviewed again by the Senate in 2010.

There is also pending GHG regulation. On December 15, 2009, EPA issued its final "endangerment finding," concluding that GHG emissions do indeed endanger public health and the environment. As a result, EPA is prepared to issue regulations actually controlling GHG emissions for the first time. Because these new rules will actually control GHG emissions, rather than simply require GHG emissions to be monitored, these new rules will automatically trigger application of the CAA prevention of significant deterioration (PSD) and Title V operating permit programs to major stationary sources of GHG emissions.

The PSD rules include the expensive demonstration, "Best Available Control Technology," or BACT demonstration. The triggering thresholds for PSD and Title V are 100 and 250 tons per year, which is quite low when considering emissions of GHGs. Thus, in September 2009, EPA issued a proposed "tailoring" rule explaining the legal mechanism upon which it would rely to raise the "major stationary source" air pollutant emission threshold of 250 tons per year and 100 tons per year to 25,000 tons per year. With its proposed "tailoring rule," EPA also explained the manner in which it would implement the CAA permitting programs to major stationary source GHG emission sources. Currently, EPA is expected to promulgate its final rules regulating and controlling GHG emissions from light duty vehicles, as well as its final "tailoring rule," in March 2010. Though subject to legal challenge, EPA's rules promulgated at the time of this writing are currently final and effective, and will remain so unless successfully challenged directly in court, or unless Congress adopts legislation preempting EPA's regulatory authority to address GHGs under the Clean Air Act.

In addition to the legislation and regulation, there is also litigation. The case that hits most closely to home most AICHE members is probably *Comer v. Murphy Oil*, where the Fifth Circuit Court of Appeals held that plaintiffs can go to trial against fossil fuel producers to prove that the GHG produced from their production and ultimate use of fossil fuels resulted in GHG emissions which increased the strength of Hurrican Katrina, and as a result, exacerbated property damage to their property from Hurricane Katrina. The plaintiffs are seeking money damages as compensation.

Beyond enforcement scrutiny from compliance with the GHG reporting rule itself, the GHG reporting rule will cause the public release of GHG emissions data at a time when this data could be used to conduct PSD and BACT review, as well as increase scrutiny by GHG plaintiffs' attorneys.

Perhaps there is safety in numbers.

Bibliography

Mandatory Reporting of Greenhouse Gases, 74 Fed. Reg. 56,260 (Oct. 30, 2009).

Comer v. Murphy Oil USA, 2009 U.S. App. LEXIS 22774 (5th Cir. Oct. 16, 2009).

U.S.EPA Special Provisions for 2010 Fact Sheet, available at:
http://www.epa.gov/climatechange/emissions/downloads/2010SpecialProvisionsInfoSheet.pdf

U.S. EPA Frequently Asked Questions, available at:
http://www.epa.gov/climatechange/emissions/ghg_faq.html

U.S. EPA Final Monitoring Rule Overview, available at:
http://www.epa.gov/climatechange/emissions/downloads/FinalMRROverview.pdf

(43b) Workshop: CO2e Emission Reporting in Ethylene Plants - Part II

David M. Winfrey, McAfee & Taft, Missouri City, TX

Kristine D. Baranski, Cardinal Engineering, Oklahoma City, OK

Mary Ellen Ternes, McAfee & Taft, Oklahoma City, OK

Deanne Hughes, Cardinal Engineering, Oklahoma City, OK

This is a continuation of the above Workshop.
See more of this Session: Workshop: CO2 Emission Reporting in Ethylene Plants
See more of this Group/Topical: Topical 4: The 22nd Ethylene Producers' Conference

(60b) Novel Polymer Membrane Process for Pre-Combustion CO2 Capture From Coal-Fired Syngas

Timothy C. Merkel, Membrane Technology and Research, Inc., Menlo Park, CA

Meijuan Zhou, Membrane Technology and Research, Inc., Menlo Park, CA

Sylvie Thomas, Membrane Technology and Research, Inc., Menlo Park, CA

Haiqing Lin, Membrane Technology and Research, Inc., Menlo Park, CA

Adrain Serbanescu, Membrane Technology and Research, Inc., Menlo Park, CA

Gasification of coal – a low-cost and abundant domestic feedstock – and subsequent conversion to hydrogen and energy with simultaneous CO2 capture can improve national energy security in an environmentally responsible manner. To fulfill the promise of clean energy production through gasification of domestic coal, more efficient and less costly methods of separating the post-shift syngas (hydrogen and carbon dioxide) are needed. Current absorption technologies (Rectisol, Selexol, and PSA) used for this separation are costly, energy intensive, and unlikely to meet DOE CO2 capture targets.

Membranes of various types are currently being considered for H2/CO2 separations in future IGCC power plants. Key membrane challenges identified from previous work in this area include high membrane cost, low flux, instability in a challenging thermal and chemical environment, device reproducibility, scale-up issues, and poor membrane system design understanding. For an Integrated Gasification Combined Cycle (IGCC) membrane process to be successfully implemented, innovations in the membranes and membrane process designs that address these issues are required.

In this talk, novel high-temperature-stable polymer membranes with H2/CO2 selectivities of >10 and hydrogen permeances of >200 gpu at syngas cleanup temperatures of 100–200°C will be discussed. These membranes can be used in an efficient membrane process that will capture >90% of CO2 in post-shift syngas and generate a hydrogen stream with >93% purity at low cost. Results from a recent field test treating real coal-derived syngas will be presented.

Keywords: syngas, membrane, carbon dioxide, hydrogen and IGCC

See more of this Session: Carbon Capture/Separation From Gasification-Based Systems

See more of this Group/Topical: Topical 2: Advanced Fossil Energy Utilization

(60c) Evaluation of the Technical and Economic Performance of the Calcium Looping Process for CO_2 and H_2s Capture From Coal-Derived Synthesis Gas

Daniel P. Connell, Research & Development, CONSOL Energy Inc., South Park, PA

Shwetha Ramkumar, Chemical and Biomolecular Engineering, The Ohio State University, Columbus, OH

L. S. Fan, Chemical and Biomolecular Engineering, The Ohio State University, Columbus, OH

Robert M. Statnick, Clear Skies Consulting, Cornelius, NC

The Calcium Looping Process (CLP) is being developed to reduce the cost and increase the efficiency of producing hydrogen and electricity from coal-derived synthesis gas. It integrates the water-gas shift (WGS) reaction with in-situ carbon dioxide, sulfur, and halide removal at high temperatures in a single reactor, while eliminating the need for a WGS catalyst. In the CLP, a regenerable limestone-derived sorbent (i.e., CaO or $Ca(OH)_2$) is used to chemically absorb CO_2, sulfur compounds, and halides from the synthesis gas. The removal of CO_2 drives the water-gas shift reaction forward via Le Chatelier's principle, obviating the need for WGS catalyst and enabling the production of high-purity hydrogen. This hydrogen can be sold commercially or combusted in a hydrogen turbine to produce electricity (i.e., in an integrated gasification combined cycle plant). The spent sorbent is heated in a calciner to regenerate CaO for reuse in the process and to release a concentrated CO_2 stream that can be dried and compressed for sequestration. The regenerated CaO is hydrated before being reintroduced into the absorber to prevent deterioration of its performance. Laboratory experiments have shown that the CLP can produce >99% purity hydrogen with less than 1 ppm sulfur impurity. The effectiveness of the hydration step in maintaining the activity of the calcium sorbent also has been experimentally demonstrated.

This paper presents a preliminary techno-economic evaluation of the Calcium Looping Process as applied to a commercial-scale coal-to-hydrogen plant with co-production of electricity. The CLP-based plant is compared with a conventional coal-to-hydrogen plant that includes a water-gas shift reactor, a two-stage Selexol unit for H_2S and CO_2 removal, and a pressure swing adsorber for hydrogen purification. One significant advantage of the CLP over the conventional process is that CO_2 absorption occurs at high temperature (e.g., 600-700 °C), allowing the exothermic heat of reaction to be recovered for use in generating steam. Excess heat from the calciner also contributes to steam production in the CLP, as does the exothermic heat of the lime hydration reaction. As a result, the CLP plant produces significantly more net electricity than the conventional plant for a given hydrogen production rate. CO_2 emissions from the CLP plant are essentially zero, because the tail gas from hydrogen production is combusted in oxygen in the calciner to produce water and sequestrable CO_2. Challenges associated with the development and scale-up of the CLP will be discussed, and the effects of sulfur chemistry on the performance of the process will be explored. Estimated capital costs and fixed and variable operating and maintenance costs for the CLP will also be presented. This preliminary analysis shows that the CLP has the potential to improve the performance and cost of producing hydrogen from coal.

See more of this Session: Carbon Capture/Separation From Gasification-Based Systems
See more of this Group/Topical: Topical 2: Advanced Fossil Energy Utilization

Development of Freeze Granulated Oxygen Carriers for Chemical Looping Combustion of Solid Fuels

Yi Zhang, Aurora Rubel, Kunlei Liu, and James K. Neathery

University of Kentucky, Center for Applied Energy Research, Lexington, KY 40511-8410

yizhang@caer.uky.edu

Chemical looping combustion (CLC) is proving to be a promising technology for the combustion of carboneous fuels producing electricity and a sequestration-ready concentrated CO_2 stream. Recently, more attention is being given to its application to solid fuels. The University of Kentucky, Center for Applied Energy Research (CAER) is presently developing an integrated process to apply CLC to coal combustion. A first step in this work is the development of suitable oxygen carriers (OCs) specifically for the application of CLC to solid fuels.

The materials investigated in this study included Fe_2O_3, NiO and CuO on a support matrix, α-aluminum oxide. The unsupported forms of the metal oxides were examined as well for comparison. All tested OCs were produced by freeze granulation (FG) with subsequent freeze drying and high temperature calcination. After calcination at 1400°C for 6 hours, the particle size distributions were evaluated. Reaction rates were examined by a thermal gravimetric analyzer- mass spectrometer (TG-MS). A mixture of $H_2/CO/CO_2/Ar$ was used as the reducing gas and O_2/Ar was the oxidizing gas mixture. Oxidation/reduction cycles of the OCs were carried out in wet gases to simulate actual conditions in a CLC reactor while experiments from dry gases were also conducted for comparison. Reaction rates, the extent of oxidation/reduction, recycle stability, and the effect of water vapor will be discussed. The data show that the supported OCs on α-alumina all have fewer tendencies to agglomerate after multiple oxidation-reduction cycles than unsupported OCs.

Introduction

Chemical looping combustion (CLC) is a promising technology [1-4] being investigated at the University of Kentucky Center for Applied Energy Research (CAER) for the combustion of solid fuels. CLC's potential is two-fold in that it can provide a high-purity, sequestration-ready stream of CO_2 and theoretically provide a substantial improvement in thermal conversion efficiency for fossil-fuel utilization with CO_2 capture. CLC uses a solid oxygen carrier (OC), usually a metal oxide to transfer the oxygen from the air (Air Reactor) to the fuel in a reducing reactor (Fuel Reactor) without the direct contact between the fuel and air. The oxygen-depleted OC is recycled by using steam or flue gas in the loopseal to the Air Reactor where the carrier is oxidized to its original state. In CLC, air does not directly meet the fuel, and so the product stream from the Fuel Reactor is not diluted with the nitrogen in air. Then nearly pure CO_2 is obtained following condensation of H_2O. The resulting high concentration CO_2 stream is ready for subsequent sequestration without any additional purification [5]. This technology was first proposed primarily for the combustion of gaseous fuels for natural gas turbines and only recently been considered for solid fuels such as coal without gasification of the coal first to syngas [6-9].

Lately, there has been a considerable interest in the application of CLC directly to solid fuels [8, 10-17]. The main challenge with the ability to combust solids in CLC is to have an appropriate OC with fast reactivity, durability, easily separated from the fuel ash, and resistant to any detrimental effect related to exposure to ash. Several OCs have been studied for use with solid fuels [8, 12-14, 17-20]. Previous work at CAER indicated that iron oxide-based powders and catalysts were the best candidates for use as OCs for CLC of solid fuels from the OCs tested. These solids did not agglomerate; are relatively inexpensive; retained reasonable activity; and remained durable through multiple oxidation – reduction cycles. The data suggested that accessibility of reactant gases to the active sites of the OC through adequate porosity was important. It was found that increased H_2, CO concentrations resulted in increased reduction rate for the iron oxides which may prove useful in a CLC process for solid materials as it provides a method to control the reduction rate during processing [8]. Fe_2O_3 powder was used to directly combust a beneficiated gasification char which contained 60% carbon. A mixture of Fe_2O_3: char (90:10) heated at 950°C in Argon resulted in 88% removal of the carbon with the OC as the only source of oxygen. This was encouraging for the application of CLC to the combustion of solid fuels [9]. Iron oxides were also found unaffected by ash concentration up to 75% and in fact some components of the ash provided additional OC activity [21].

This work was to begin development of an appropriate OC for the CLC of a devolatilized coal char in a bench scale fluidized bed (FB) CLC Redox reactor being constructed at the CAER. Freeze granulation was chosen to develop an OC composed of a metal oxide on a solid support of alumina which had suitable reactivity, particle size, porosity, redox recycle ability, and durability for use in the FB reactor. Three metal oxides were chosen for this study, Fe_2O_3, NiO, and CuO. Fe_2O_3 was our main interest with NiO and CuO used for comparison. To help determine the best metal oxide to alumina ratios, the extent and rates of oxidation/reduction, the redox recycle ability, and the effect of moisture were studied by TG -MS.

Experimental

Commercially available metal oxides (Fe_2O_3, NiO, and CuO) and α-alumina (Al_2O_3) powders were used. A slurry, referred to as a slip in this paper, was prepared by ball milling a mixture of solids, additives (dispersant and binder) with deionized water. The slip was pumped via a peristaltic pump to the spray nozzle of the freeze granulator (Figure 1). Pressurized air was also sent to the spray nozzle to combine with the slip creating a uniform spray mist which is sent into a magnetically stirred container of liquid N_2. The spray is regulated by the air pressure and the feed pumping rate of the slip. This results in the production of frozen spherical particles. The size of the spheres (OCs) can be adjusted by changing the air pressure, the feed rate, and composition of the slip. The metal oxide to alumina loadings used to produce freeze granulated (FG) OCs were 100:0, 75:25, 50:50, 25:75 of wt/wt % for each of the three metal oxides studied. Freeze granulation conditions, pump speed and air pressure, were optimized to maximize the particle yield in size range desired. The slips were prepared by ball milling for at least 24 hr. Air pressure was adjusted in the range of 40~60 kPa. The slip flow rate was 23.4~40 ml/min. FG particles were kept in the frozen state until freeze dried. After freeze drying, the FG OCs were calcined at 1400°C for 6 hours to achieve high durability. For CuO samples, the calcination temperature was set at 800 °C. The same freeze drying and calcination procedures were used for all samples. Samples were sieved to 48x150 mesh particle size samples which were tested for extent and rate of oxidation/reduction, redox recycle capability, and agglomeration resistance by thermal analysis-mass spectrometry (TG-MS) under both dry and wet gases.

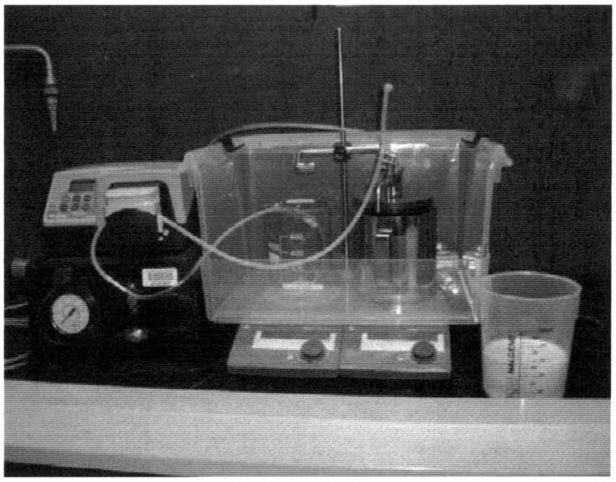

Figure 1. Freeze granulator equipment.

The TG-MS was a Netzsch STA 449 F3 Jupiter equipped with a water vapor furnace and coupled to a Aeolos QMS403 (Figure 2). It is capable of temperature to 1200°C under both steam and dry gases. The MS measured gases leaving the TG to determine the gas constituent in the evolved stream from TG. Conditions for testing are reducing gases of 10% H_2, 15% CO, 20% CO_2 with and without steam and a balance of Ar; oxidizing gases of 20% O_2 in a balance of Ar; and a reaction temperature of 950°C. Oxidations/reductions were carried out to near completion and at least 3-5 redox cycles were done per OC. Gas flow rates and switching was controlled by a set of mass flow controller operated by Lab View software on a dedicated computer. Times for oxidation/reduction and flows could be programmed as needed and the entire operation could be run 24/7 without intervention.

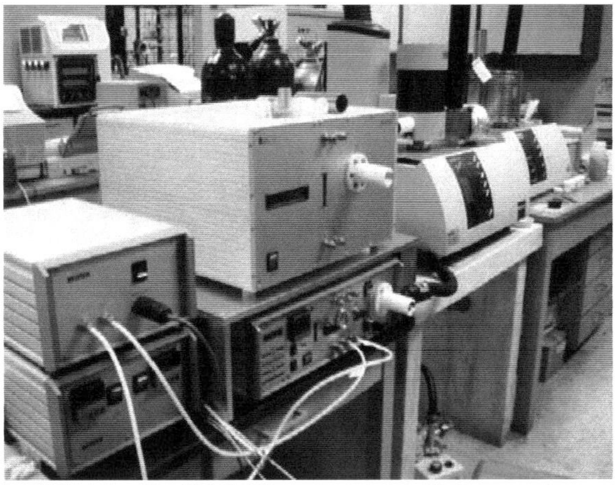

Figure 2. TGMS system used for study.

Results

Figures 3-5 give the weight gain/loss on oxidation/reduction for the different metal oxides loading on alumina for Fe_2O_3, NiO, and CuO respectively. FG Fe_2O_3 showed a slight decrease in the amount of oxygen released (active-O) from the OC after dilution correction as the active metal oxide content decreased. Under wet gases, there was a significant decrease in the weight gain/loss for unsupported (100%) FG Fe_2O_3 OC which did not occur for supported OCs. As contrary, the same amount of oxygen was released during oxidation/reduction of the Fe_2O_3/alumina OCs as in dry gases (Figure 3). The unsupported (100%) FG NiO OC showed a slight decrease in active-O release during the five cycles. However, in previous studies [8, 9] reagent-grade NiO powder agglomerated in one cycle. The unsupported FG NiO OC performance was far better than the pure powder. There was little difference in the weight gain/loss on oxidation/reduction with metal loadings of NiO when run in dry gases. Under wet conditions, the 75% NiO on alumina corrected for dilution gained and lost more active-O than lower loadings (Figure 4). Unsupported FG CuO OC (Figure 5) when run in dry gases agglomerated while the support CuO OC remained active through multiple cycles at 950°C which is substantially higher than the 800°C used for pure CuO powder in a previous work which became inactive quickly due to agglomeration. . There was a decrease in the weight gain/loss on redox for all CuO OCs in wet gases compared to redox in dry gases but there was no evidence of agglomeration or loss of activity with cycles.

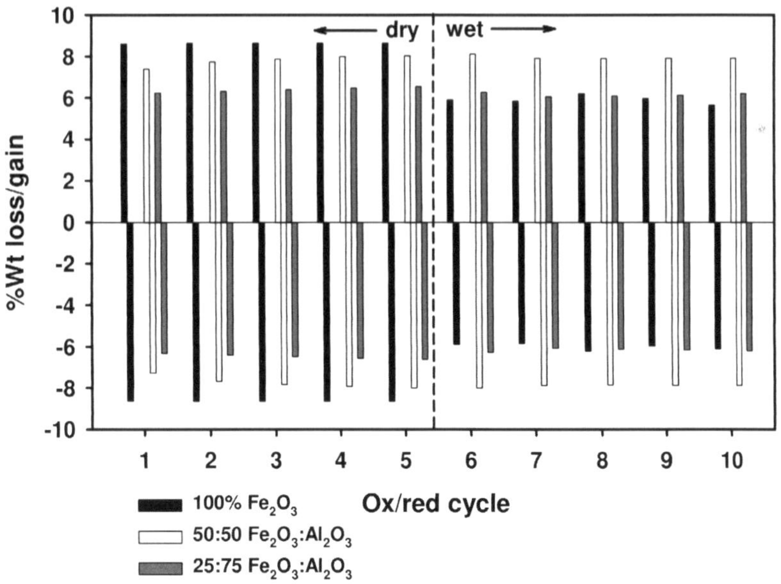

Figure 3. Weight gain/loss on oxidation/reduction of FG Fe_2O_3.

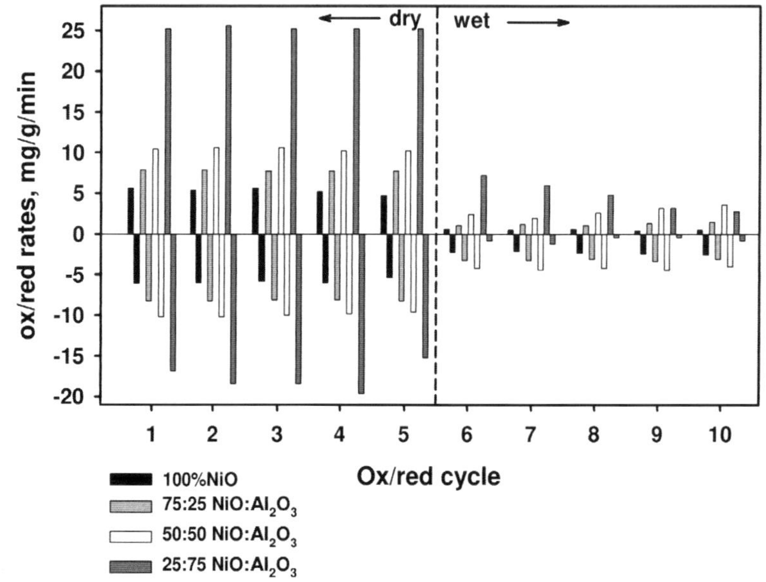

Figure 4. Weight gain/loss on oxidation/reduction of FG NiO.

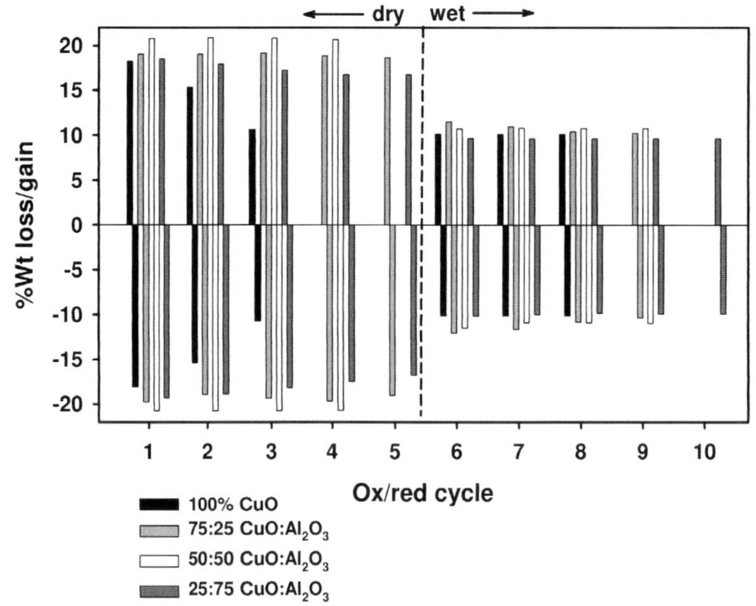

Figure 5. Weight gain/loss on oxidation/reduction of FG CuO.

The loading of active metal oxides on alumina by FG resulted in a significant effect on the rates of oxidation/reduction. For FG Fe_2O_3 cycled between oxidation/reduction under dry gases, there was an increase in the rates as the active metal loading decreased (Figure 6). Under wet conditions rates decreased from comparable dry condition but again there appeared to be a trend toward faster rates with low metal loading especially during reduction though there was considerable variability. One interesting effect of the FG process was that the rates of reduction were greater than oxidation which was different from results with pure metal oxide powders where oxidation rates far exceeded reduction rates [8, 9]. Figure 7 gives the results for the rates of oxidation/reduction for NiO. The trends under dry gases for NiO were the same as for Fe_2O_3 except the increase with lower metal loading was much greater. The increases for supported FG Fe_2O_3 and NiO were 2 and 5 times their unsupported FG OCs respectively. Under wet gas conditions, rates decreased greatly over dry gas conditions, 5 times slower rates for all loadings of NiO. CuO rates (Figure 8) had the same trend as for the other two metal oxides with moist gases even more detrimental with oxidation rates near zero.

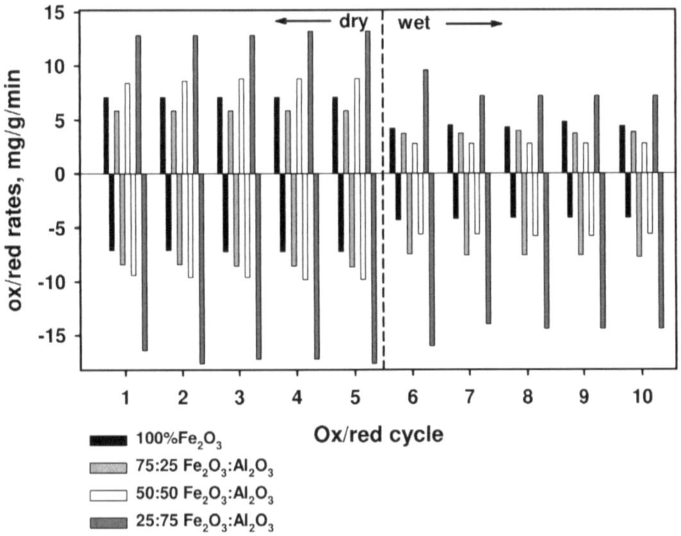

Figure 6. Rate of oxidation/reduction for FG Fe_2O_3 under dry and wet gases.

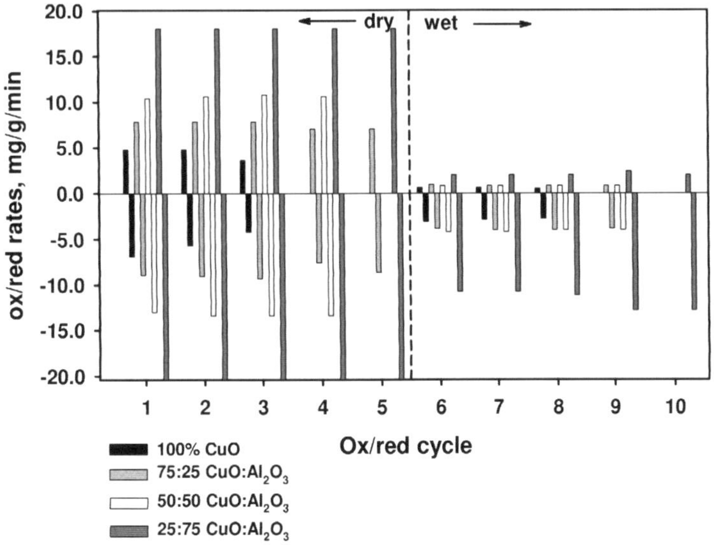

Figure 7. Rates of oxidation/reduction for FG NiO under dry and wet gases.

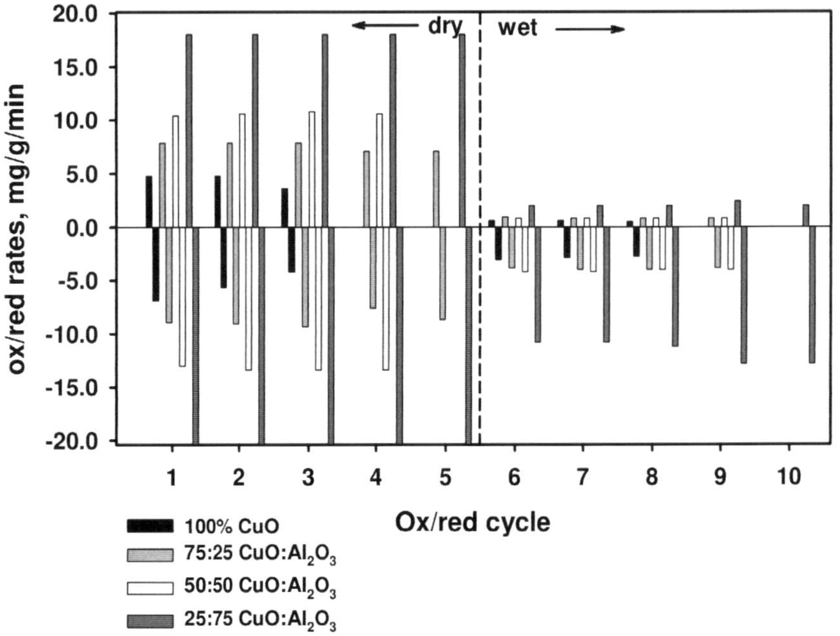

Figure 8. Rates of oxidation/reduction for FG CuO under dry and wet gases.

<u>Discussion</u>

The results suggest the complexity of factors affecting the FG OCs produced in this study. As others have observed loading active metal oxides on a support is beneficial in preventing agglomeration. Our previous work with pure NiO powders [8, 9] agglomerated on the first redox cycle and were no long active after the first cycle. The loading of CuO on a support in this study allowed for the use of CuO at a temperature significantly above that which the pure CuO powders could be used. In this study, CuO/alumina OCs remained active at 950°C where the unsupported (100%) FG CuO OC lost activity due to agglomeration or melting. The spent unsupported FG CuO was a pellet melt of what appeared to be pure Cu.

For all metal oxides in dry gases, the oxidation/reduction rates increased with lowering the active metal oxide concentration in the supported OCs. The definitive explanation for this will undoubtedly involve some combination of pore size, access of gases to reactive sites, spatial proximity of active sites and particle size. All issues which are currently being investigated at the CAER. Each metal oxide will form a spatial arrangement on the alumina support which will depend on its granular size and loading. The reaction rates for the FG metals was significantly lower than for the pure metal oxides again due to limitation of accessibility of redox gases to active oxygen sites which are most likely impacted by particle size, geometry, and accessibility . However, this is probably what protects Ni and Cu from agglomerating as the slower rates reduces temperature extremes and individual metal molecules may not be in close proximity to agglomerate. The slow reaction rates for all the metal oxides in the presence of moisture especially for NiO and CuO is difficult to explain other than the possible blocking of active sites. Water has been shown to react with alumina [22] which could lead to alteration of the geometry and reactivity of the FG OC. However, the better reduction rates over oxidation rates in moist gases could be due to an increase in H_2 via the water gas shift reaction since previous studies on pure metal oxides found that increasing the H_2 content of the reducing gases increased the reduction rate [8, 9].

Conclusions

Though there were benefits (no agglomeration and melting) to the FG of NiO and CuO, the slow rates of oxidation/reduction do not make them viable OC for the CLC of solid fuels. As with our previous studies Fe_2O_3 would be the OC of choice. FG Fe_2O_3 had similar weight gains and losses on oxidation/reduction, were not affected by wet gases when supported on alumina, and maintained better reaction rates in moist gases compare to NiO and CuO. The advantage of FG is that the OC particle size and metal loading can be adjusted to suit the necessary parameters required for separation and reactivity in a CLC reactor.

Acknowledgement

Funding for this work was provided by the Carbon Management Research Group.

References

[1] Croiset, E Thambimuthu, K. Coal Combustion with Flue Gas Recirculation for CO_2 Recovery. In: Riemer, P, Eliasson, BWokauun, A, Editors. Greenhouse Gas Control Technologies, New York: Elsevier; 1999, p. 581-6.
[2] Marion, J, Naskala, Y N, Griffin, T. Controlling Power Plant CO_2 Emissions: a Long Range Approach. In: The First National Conference on Carbon Sequestration, Washington, DC,2001; DOE/NETL-2001/1114
[3] Herzog, H, Eliasson, B, Kaarstad, O. Capturing Greenhouse Gases. Scientific American 2000; 282(2):72-9.
[4] Hatanaka, T, Matsuda, S, Hatano, H. A New-concept Gas–solid Combustion System 'MERIT' for High Combustion Efficiency and Low Emissions. In: Intersociety Energy Conversion Engineering Conference 1997:944-8.
[5] Freund, P. Abatement and Mitigation of Carbon Dioxide Emissions from Power Generation. Powergen 98 Conference. Milan,1998; Powergen 98 Conference
[6] Richter, H Knoche, K. Reversibility of Combustion Processes. In: ACS,1983:71-86.
[7] Anthony, E J. Solid looping cycles: A new technology for coal conversion. Industrial and Engineering Chemistry Research 2008; 47(6):1747-1754.
[8] Rubel, A, Liu, K, Neathery, J, Taulbee, D. Evaluation of oxygen carriers for chemical looping combustion of coal and solid fuels. In: 24th Pittsburgh Coal Conference, Johannesburg, SA,2007:paper 21-1.
[9] Rubel, A, Liu, K, Neathery, J, Taulbee, D. Oxygen carriers for chemical looping combustion of solid fuels. Fuel 2009; 88(5):876-884.
[10] Cao, Y, Casenas, B, Pan, W P. Investigation of chemical looping combustion by solid fuels. 2. Redox reaction kinetics and product characterization with coal, biomass, and solid waste as solid fuels and CuO as an oxygen carrier. Energy and Fuels 2006; 20(5):1845-54.
[11] Cao, Y Pan, W P. Investigation of chemical looping combustion by solid fuels. 1. Process analysis. Energy and Fuels 2006; 20(5):1836-44.

[12] Chuang, S Y, Dennis, J S, Hayhurst, A N, Scott, S A. Development and performance of Cu-based oxygen carriers for chemical-looping combustion. Combustion and Flame 2008; 154(1-2):109-121.
[13] Leion, H, Jerndal, E, Steenari, B-M, Hermansson, S, Israelsson, M, Jansson, E, Johnsson, M, Thunberg, R, Vadenbo, A, Mattisson, T, Lyngfelt, A. Solid fuels in chemical-looping combustion using oxide scale and unprocessed iron ore as oxygen carriers. Fuel 2009; 88(10):1945-1954.
[14] Leion, H, Lyngfelt, A, Mattisson, T. Solid fuels in chemical-looping combustion using a NiO-based oxygen carrier. Chemical Engineering Research and Design 2009; 87(11):1543-1550.
[15] Leion, H, Mattisson, T, Lyngfelt, A. CO_2 capture from direct combustion of solid fuels with Chemical-Looping Combustion. In: Proceedings of the 33rd International Technical Conference on Coal Utilization and Fuel Systems, Clearwater, FL, USA,2008
[16] Leion, H, Mattisson, T, Lyngfelt, A. Combustion of a German lignite using chemical-looping with oxygen uncoupling (CLOU). In: The Proceedings of 33rd International Technical Conference on Coal Utilization & Fuel Systems, Clearwater, FL, USA,2008
[17] Siriwardane, R, Tian, H, Richards, G, Simonyi, T, Poston, J. Chemical-Looping Combustion of Coal with Metal Oxide Oxygen Carriers. Energy & Fuels 2009; 23(8):3885-3892.
[18] Mattisson, T, Leion, H, Lyngfelt, A. Chemical-looping with oxygen uncoupling using CuO/ZrO2 with petroleum coke. Fuel 2009; 88(4):683-690.
[19] Zhao, H, Liming, L, Xu, D, Chuguang, Z. Study on Ni-based Oxygen Carriers for Chemical Looping Combustion by Coal Char. In: 24th International Pittsburgh Coal Conference,2007:1-18.
[20] Zhao, H B, Liu, L M, Wang, B W, Xu, D, Jiang, L L, Zheng, C G. Sol-gel derived $NiO/NiAl_2O_4$ oxygen carriers for chemical-looping combustion by coal char. Energy and Fuels 2008; 22(2):898-905.
[21] Rubel, A, Liu, K, Neathery, J. The Effect of Ash on Oxygen Carriers for Chemical Looping Combustion of Solid Fuels. In: AIChE Extended Abstracts, Nashville, TN,2009
[22] Wittbrodt, J M, Hase, W L, Schlegel, H B. Ab Initio Study of the Interaction of Water with Cluster Models of the Aluminum Terminated (0001) a-Aluminum Oxide Surface. Journal of Physical Chemistry B 1998; 102(34):6539-6548.

(60e) Pulverized Coal Pyrolysis & Gasification in $N_2/O_2/CO_2$ Atmospheres by Thermo-Gravimetric Analysis

Muhammad Faisal Irfan, Living and Environmental Sciences, Fukuoka Women's University, Fukuoka, Japan

The combustion characteristics of Datong coal in the different $N_2/O_2/CO_2$ environments have been extensively studied by using thermo-gravimetric technique. The effects of combustion environment, oxygen and CO_2 concentrations, pyrolysis conditions, particle size and different heating rates were considered. Results showed that, at each temperature zone, the mass loss rate of coal pyrolysis in CO_2 environment is higher than that in N_2 environment. This may be due to the division of CO_2 pyrolysis process into three stages: moisture release, devolatilization and char gasification by CO_2 at higher temperature zone or due to the formation of perfectly hexagonal structure of coal. However, the burning process of coal in O_2/CO_2 environment is almost same compared with that in O_2/N_2 environment for equivalent oxygen concentrations but slightly lesser than O_2/N_2 environment at the end of the process. With the increase of oxygen concentration, the burning rate of coal increases and burnout time is shortened. Among the different pyrolysis temperatures, 500 oC was found to be the best temperature as the ignition point in the O_2/CO_2 atmosphere lies near or just after this temperature. Moreover, the surface area produced by this char at this temperature was much higher in CO_2 atmosphere (400 m2/g) than N_2 (52 m2/g). This simultaneous process of pyrolysis and gasification seems to be advantageous because of production of the nascent char in one process and avoiding of the excess use of gasifying medium from room temperature to 500 oC. As the heating rate increases, coal particles are faster heated in a short period of time and burnt in a higher temperature region, but the increase in heating rate has almost no obvious effect on the combustion mechanism of coal. At different heating rates the values of the Arrhenius activation energy (E) and pre-exponentional factor (A) were estimated from the TGA data by applying different methods such as Freeman-Carroll, Doyle and Coats-Redfern. A trend of decrease of activation energy values were observed with the increase of different heating rates.

See more of this Session: Carbon Capture/Separation From Gasification-Based Systems

See more of this Group/Topical: Topical 2: Advanced Fossil Energy Utilization

FUTURE FEED-STOCKS FOR PETROLEUM AND PETROCHEMICAL INDUSTRIES

Dr D P Lal

1. Indroduction

. The phenomenal rise in crude oil price, observed sometime back, has increased the urgency to explore the alternate feed-stocks which the petroleum and petrochemical industries have to adopt in future. Though there is a temporary drop in price, but once the world economic conditions become normal, such up-ward price trend, cannot be ruled out. This is due to the fact that the world crude oil reserve explored so far (approx. 1230 billion barrels) is expected to last only 70-80 years, current consumption level being approx. 30 million barrels per day. Even if the reserve under the arctic (which is estimated at around 90 billion barrels) is explored, this additional reserve will meet world demand for another 9 years only. Further, it is estimated that there will be 35% increase in worldwide petroleum demand by the year 2015. This rising crude oil demand against falling reserve would eventually push up the crude oil price to uneconomic level. So the world is faced with the daunting question;- how this rising demand is going to be met economically?

2. Need to Search for Alternate Feed-stocks

For reasons mentioned above, it has now become imperative to find out alternate feed-stocks for petroleum and petrochemical industries on a priority basis. It is predicted that after 20-30 years, the price of many of the alternate feed-stocks now being explored and described later in the paper, would become competitive compared to price of crude oil based feed-stocks. Hence, most of the developed countries in the world, as well as many developing countries are now engaged in this phenomenal task of finding out appropriate alternate feed-stocks.

The other compelling factor which has forced the world community to look for alternate feed-stocks and fuels is the harmful contribution of the fossil fuels to global warming. While the present international search is mainly aimed at finding alternate resources to meet the energy requirement of the world without contributing to or preferably reducing the global warming; but the search will also lead to finding out liquid products containing linear and aromatic hydrocarbons from sources other than crude oil which could be recovered to produce feed-stocks for refineries and petrochemical plants.

In this paper, however, we will concentrate on those areas of research and development which are going to lead to finding out alternate feed-stocks for refineries and petrochemical plants.

3. New Feed-stocks in the Initial Phase

It need not be emphasized that techno-economic criteria will ultimately determine how different feed-stocks are going to be adopted in the concerned industries as the time passes. Hence, one of the deciding criteria for selecting the new feed-stocks will be how best and how long the existing investments made in this field (namely the existing petroleum refineries) can be utilized with modifications wherever necessary. From this consideration; it becomes obvious that residues, heavy crude oil and high sulphur crude oil which are presently relatively uneconomic to process, will be the first group of feed-stocks to be considered. The technologies required for processing these feed-stocks (namely, hydro-cracking process, coker technology, sulphur removal process etc.) are well-proven and are already in operation in certain refineries and can be easily adopted in other refineries. But profitable and environmentally safe outlets of the by-products produced in bulk scale from these feed-stocks (i.e petroleum coke and sulphur) should be found out before these feed-stocks are adopted in any refinery.

The transportation of these feed-stocks to any refinery would not be a problem, because, the existing means of transportation (i.e LCC's and VLCC's) can be used for this purpose.

The next group of feed-stocks which could be considered for processing in the existing refineries are the oil stocks recovered from tar-sands and tar-shale. New technologies required for this oil recovery are in advanced stage of development mainly due to extensive R&D work done under the sponsorship of DOE, USA. While, number of pilot plants are in operation within USA for oil recovery from tar-sand; ENI is already planning to invest approx. US $ 3 billions to recover oil from tar-sand in Congo. From the scale of investment, it seems that this is going to be a commercial plant; which is an indication that the technology has sufficiently developed, so that a commercial scale plant could be built based on the latest advances made. However, since transportation of this raw material (i.e tar sand) over long distance would be uneconomic, hence these plants should be located near the source of tar-sand. But the oil recovered could be transported over long distance to far off refineries using conventional modes of transportation.

The liquid products produced from the refineries by processing the above-mentioned feed-stocks would be substantially similar to those produced from crude oil, available at present. And hence the present types of products (i.e, naphtha, BTX, kerosene etc) would continue to be available for the downstream petrochemical plants. Thus, no major change is anticipated in the technologies presently being used in the petrochemical plants for treating the feed-stocks supplied from refineries processing residue, heavy crude oil etc..

4. New Feed-stocks in the Second Phase

Many of the countries in the world have got huge reserves of coal and natural gas. With price of fossil fuels going up, these countries will endeavor to produce liquid hydrocarbons based on these raw materials. In fact, the world coal reserve is estimated at around 100,000 quadrillion btu; and is expected to last for approx. 300 years. Thus countries like India and Australia, which are rich in coal reserve, could try to make liquid fuels from coal, based on some of the established process. South Africa, another coal-rich country is already producing various liquid hydrocarbons from coal.

There are two process steps to convert coal to liquid hydrocarbons. At first, gasification of coal is done with a mixture of steam and air to produce synthesis gas (a mixture of hydrogen and carbon-monoxide) by using Kopper-Totzek,, Shell or BGL coal gasification process. The synthesis gas so produced is converted to liquid hydrocarbons through Fischer Tropche process. In addition, in the steel industries, coke required in blast furnace is produced through pyrolysis of raw coal, which releases number of aromatic hydrocarbons (benzene, toluene, xylene, napthene etc.) which could be recovered in pure form and used as feed-stocks of chemical and petrochemical industries.

The Sasol coal-based complex in South Africa is the fore-runner in this field. Besides fertilizers and fuels, Sasol is producing numerous liquid hydrocarbons from coal, which are used as feed-stocks in various chemical and petrochemical industries. Syntroleum, a USA company is producing jet-fuel and diesel from synthesis gas. There is some initiative to set up coal based liquid hydrocarbon production plants in India in collaboration with Sasol.

The interesting feature of the Fischer-Tropsch process is that the product profile can be adjusted by using appropriate type of catalyst. In fact, a patent has been taken recently where the catalyst can yield olefinic naphtha and hydrogenated distillate suitable for use as naphtha cracker feed-stock.

Like coal, natural gas is likely to be an important feed-stock for production of liquid hydrocarbons. The world natural gas reserve is estimated at around 6100 trillion cubic feet and is expected to last for around 250 years. Approx. 40% of the reserve lies in the Middle East and 25% in Russia. Rest of the reserve is spread over mostly in Asia and North & South Americas. Conversion of natural gas to synthesis gas is done through steam reforming process widely used in the fertilizer plants all over the world. Thereafter, conversion of synthesis gas to various hydrocarbons could be done using the Fischer-Tropsch process or its modified versions. Shell has developed successfully a very clean and efficient process to produce diesel from natural gas.

At this point, it is worth mentioning that compositions of liquid products produced from synthesis gas as discussed above would be substantially different from those of fossil fuels. Hence, the scope for utilizing the existing refineries to recover selected cuts from the liquid products produced in Fischer-Tropsch process based plants, would be rather limited. It would be a chemical engineering challenge at that point of time as how best to utilize the investments made in the existing refineries to achieve this objective. However, feed-socks for the petrochemical plants as used presently would be available from the Fischer-tropsch process based plants and hence the existing petrochemical plants would remain suitable to treat new feed-stocks. This is obvious from the experience of Sasol.

5. Renewable Feed-stocks like Bio-fuels

The general consensus of the experts is that bio-fuels are the answers to meet the energy need of the future generations, because of the duel benefits offered by this type of fuels. These benefits are;- 1) the feed-stocks for bio-fuels are renewable in nature and would be available indefinitely unlike fossil fuels, coal and natural gas which have limited reserves; and 2) there is no net release of carbon-di-oxide when bio-fuels are used, because during growth of the plants supplying raw materials for production of the bio-fuels, nearly as much carbon-di-oxide is consumed by the plants, as is released during combustion of the corresponding quantity of bio-fuels. Hence, it is an elegant solution to the problem of global warming by limiting release of carbon-di-oxide to the atmosphere.

Extensive research and development work has been going on in different countries all over the world for finding out the most appropriate process for production of bio-fuels which will not only meet the fuel requirement and arrest the global warming, but also will not have any adverse impact on food crops production. The research work and field studies done with these objectives have already led to development of three generations of bio-fuels, which are reviewed in the following paragraphs.

In fact most of the developed and developing countries have made it obligatory to blend fossil fuels with bio-fuels (Renewable Transport Fuel Obligation – RTFO). Consequently, world bio-fuel demand has approached nearly 100 million tonnes per year. Hence, well known oil companies like Shell, Chevron, BP, ENI, Petrobas, Virgin Atlantic etc. are also deeply involved in developing bio-fuels from various feed-stocks.

5.1 First Generation of Bio-fuels

The first generation of bio-fuel started with production of alcohols from sugarcane and corn. In this development, Brazil took the lead with extensive sugarcane cultivation followed by USA producing alcohol from corn. Brazil now generates nearly 47% of its electricity requirement from renewable resources;-mainly alcohol and hydro-electricity.

Dry land crop sorghum is also an attractive crop for alcohol production, because the juice extracted from it contains 15-20% sugar compared to 10-15% in that extracted from sugarcane; and comparatively much less water is required for cultivation of sorghum.

But there are some major adverse impacts of producing bio-fuel like ethanol from sugarcane and corn. Because, diversion of land area from food crop cultivation to corn and sugarcane cultivation is leading to scarcity of land area available for growing food crops and hence rise in food crop price. According to a study by IMF, approximately 10% rise in food crop price is attributable to this factor. In addition, heavy use of fertilizers and pesticides for cultivation of these crops pollutes ground and river water. The extensive deforestation which is taking place particularly in Amazon rain forest area and Bolivia is attributed to this diversion of land area for sugarcane and corn cultivation. This is also leading to global warming because of reduction in carbon-di-oxide sinks. Because of these adverse impacts; USA, Japan and the European countries are discouraging any further blending of alcohol with petrol. However, this is not applicable to all countries. In any country, if waste dry land is available, sorghum could be grown there to produce alcohol without any adverse impact on food production.

The other type of bio-fuel, namely bio-diesel is being developed and produced in semi-commercial scale in different countries from wide variety of feed-stocks, i.e, palm oil, coconut oil, castor oil, soybean oil, jatropha seed oil etc. It may be of interest to know that Rudolf Diesel, the inventor of diesel engine first tried out his engine with peanut oil and he expected that vegetable oils would be used for driving diesel engines. This, however, did not materialize because of wide availability of diesel from petroleum refineries. But the financial viability of production of bio-diesel from feed-stocks like palm oil, coconut oil, castor oil etc. depends to a great extent on the price of the feed-stocks, which is quite volatile. Depending upon international availability, the price of any of these feed-stocks could be quite low or un-economically high. Hence for long term economic viability of these diesel production units; these should be so designed that these are capable of operation with any of these feed-stocks without sacrificing operating efficiency.

Asiatic Centre of Genome Technology in Malaysia has completed sequencing study of palm genome and jatropha genome to develop high oil yield varieties of these plants. Malaysia and Indonesia are targeting to replace 5% of their fuel consumption by bio-diesel within 2-3 years. India is targeting to replace 10% of transport fuel by bio-fuel by 2017.

A study funded by DOE, USA states that enough bio-diesel could be produced to meet total demand of diesel in USA (approx. 140 billion gallons) from oil growing plants planted over a land area of 9.5 million acres which is only a small fraction of total crop land area (approx. 450 million acres) now being used in USA. There are more than 20 companies in USA producing bio-fuels (both alcohol and bio-diesel) to meet the obligatory blending requirement of petrol and diesel with respective bio-fuels.

A British Company, Sun Bio-fuels PLC has invested $20 million in Tanzania to plant jatropha oil seeds over a land area of 9000 hectares to produce bio--diesel. Besides bio-diesel, they intend to produce glycerine (a by-product of the process for recovery and purification of bio-diesel from jatropha seed oil), varnishes, soaps, pesticides and medicines for skin disease. A Chennai (India) based Company, DI Mohan Biofuels Ltd is planning to invest $ 17 million to process 24000 tonnes of jatropha seeds annually to produce 8000 tonnes of bio-diesel along with by-products like glycerne, varnish etc. Indian Railway is considering cultivation of jatropha over 90,000 acres of land area available to them, to meet their diesel requirement.

Countries like India which have large area of dry waste land are in a unique position to produce bio-diesel from jatropha seeds. Because, these drought tolerant plants can be grown on dry wasteland. And hence cultivation of jatropha would not compete with production of food crops. If whole of 60 million hectares of wasteland available in India is used for jatropha cultivation, then approx. 30 million tonnes of bio-diesel could be produced from the jatropha seeds, besides large volumes of other by-products like glycerine, soap, varnish etc. It goes without saying that this would give a big boost to rural industrialization and employment. Such large area covered with green plants would act as a large sink of carbon-di-oxide and contribute to arresting global warming.

It is worth-mentioning that Botanical Survey of India has identified 400 species of plants and trees which can yield oil. But jatropha has emerged as the preferred option. The other plants which have good potential are pangamia and ponamia pinnala. Pangamia is a leguminous tree and is helpful for nitrogen fixing. Its leaves are used as organic manure.

Relative oil yield of various types of plant products per acre of plantation is furnished in Table -1.

5.2 Second Generation of Bio-fuels

Research and development on second generation of bio-fuels started with the objective of producing this product from waste vegetable items like corn stover, cane bagasse, wheat straw, hay, soft wood, stems, leaves, wood chips, pulp, various types of grass and other similar cellulosic waste materials. Ligno-cellulosic compounds in these materials are broken down with enzymes to free the sugar molecules which are then separated and fermented to produce alcohol. Number of patents have been taken to produce ethanol from lingo-cellulosic feed-stocks mentioned above. USA, Canada, Denmark, China, Japan and Brazil have already set up semi-commercial scale cellulosic ethanol plants to produce bio-ethanol from the second generation feed-stocks

Direct conversion of lingo-cellulosic feed-stocks to gasoline like products is the latest development in this field. A research team in MIT has patented a process which involves rapid heating of this type of feed-stocks in a liquid medium in presence of certain catalyst followed by rapid cooling which yields a liquid mass containing many of the components of gasoline. It seems second generation of bio-fuels i.e cellulosic ethanol has got a distinct commercial possibility and is likely to compete with the first generation bio-fuels.

5.3 Third Generation of Bio- fuels

Third generation of bio-fuels have been developed using various types of algae as feed-stocks. There are number of advantages of using algae as raw material for production of bio-fuels. These are;-

1) Oil-yield from algae could be up to 15 times higher compared to other raw materials mentioned above per acre of land area required for growing these feed-stocks..

2) By developing appropriate culture of algae, either diesel or petrol can be produced from it, as well as the oil yield can be improved.

3) Algae can be grown either in fresh water, or in salt/brackish water, or in contaminated water/ effluent, in pond or in back-waters along sea coast. Thus algae can help treatment of effluent from a locality and there is minimum constrain regarding medium of its growth.

4) Growth of algae can be accelerated by passing carbon-di-oxide through the algae containing mass. Thus it could solve the problem of greenhouse gas emission from a power plant by absorbing carbon-di-oxide from the flue gas emitted from the power plant.

With all these significant advantages, no doubt algae has become one of the most preferred feed-stock for producing bio-fuels. Hence, extensive research and development work has been taken up in most of the developed countries and in some of the developing countries, to find out the most appropriate type of high oil yield variety of algae. Exxon, Mobil, Shell and Department of Defense (USA) are making serious efforts to be the commercial producer of bio-fuel from algae. DOE funded a project of quick growing algae for sequestering carbon-di-oxide emission from coal-fired power plant. Some of these algae are of high oil yield variety. USA is also collaborating with China for developing high oil yield variety of algae. As a result of persistent R&D work, 15 algae based bio-fuel plants are in the process of construction in USA. The largest among these is the one by Petrosun Ltd to be built in Rio Hando, Texas; to produce 4.4 million gallons of bio-fuel from algae. While most of the development so far has taken place with algae grown in ponds; but research work is in progress to grow it inside bio-reactors either exposed to sun-light or to artificial light. A Dutch Company has developed photo-bioreactors which can produce 6-times more algae even in cold climate. Arizona Public Service Company uses algae bioreactors to capture carbon-di-oxide from flue gas of their power plant. Oil recovered from algae thus grown is used as fuel in the same power plant.

The bio-reactors are getting more popular since these occupy much less space compared to large water body required for algae growth and relatively oil yield being much higher. Oil recovery process from algae is very simple. Oil is extracted by expeller followed by solvent extraction for recovery of residual oil from algae. The algae cake can be used as fuel.

The process of producing bio-fuel from algae has got very high potential. Because, there is immense scope of improving this process on various fronts. Through genetic study of algae strains, its oil yield could be increased manifold. Already some researchers have succeeded, in laboratory scale, to produce algae strains which yield nearly 10 times more oil compared to the present varieties. Through systematic bio-chemical engineering study, the design of the bio-reactors used for algae growth could be improved to make these;- high throughput, non-clogging, continuously operated ones. This would improve plant productivity and reduce plant cost; and at the same time occupy minimum land area.

In spite of numerous advantages of third generation of bio-fuels, first and second generation bio-fuels might be worth producing in some areas because of some local advantages. For example, countries producing sugar would find it advantageous to produce alcohol through fermentation of molasses, since this is a profitable way of disposing off this effluently. Similarly, countries having large areas covered by wasteland would find it beneficial to plant jatropha trees in the wasteland, thereby producing bio-diesel from jatropha seeds; and at the same time increasing green coverage thereby reducing global warming and improving climatic conditions. In the same way, where large volume of cellulosic wastes (hay, wheat straw, wood chips, soft wood etc.) are available, it might be a profitable way to dispose off these materials by producing bio-fuels from these waste products. Hence, research and development work is going on in different countries on all these three generations of process for bio-fuel production.

6. Scope of Using Bio-fuels as Feed-stocks for Petrochemical Industries

As is obvious from above description, so far only two types of bio-fuels have been produced in semi-commercial scale. These are;- bio-alcohol and bio-diesel. Only one study has been reported where it was possible to produce a mixture of hydro-carbons like fossil fuels in laboratory scale from feed-stocks used in developing second generation of bio-fuels. Techno-commercial feasibility of producing this product is yet to be proven.

However, from bio-alcohol, ethylene could be produced using the well-established alcohol de-hydration process. The process of conversion of ethylene to higher olefins like propylene, butene, butadiene etc. have been already developed. Thus, there could be a mother plant using bio-alcohol as the feed-stock and producing a range of olefins which could be used as feed-stock for various polymer plants and other olefin-based chemical plants.

Since production of benzene from ethylene is till now an expensive process, hence there could be a problem to meet economically the feed-stock requirement for the aromatics-based chemicals and polymers plants based on alcohol. However, benzene and higher aromatics should be available from coal through pyrolysis process (as is used for producing coking coal in steel plants) for another two centuries or so. By that time, hopefully it would be possible to develop aromatic products ranging from benzene to higher aromatics based on renewable natural resources as feed-stocks, as has been developed in laboratory scale now.

7. Conclusions

Based on the facts and figures presented above, it could be concluded that after reserves of fossil fuels are exhausted, a range of alternate feed-stocks would appear in the market to meet the feed-stock requirement of petroleum and petrochemical industries. Alternate feed-stocks which are likely to be used initially are heavy and high-sulphur crude oil, followed by oil recovered from tar-sand and tar-shale. Simultaneously coal-based feed-stocks would be developed particularly in the coal-rich countries, as has already happened in South Africa. Countries rich in natural gas reserve would then start marketing the liquid hydrocarbon products synthesized from natural gas following the process described earlier.

Bio-fuels like bio-alcohol and bio-diesel are already used in both developed and developing countries as partial replacement of fossil fuels. Though presently this replacement is mainly aimed at reducing greenhouse gas concentration in the atmosphere to arrest global warming and thereby improve climatic conditions; but the proportions of the bio-fuels are likely to increase further because of the price competitiveness of bio-fuels against fossil fuels and also because of the over-riding concern to control global warming. Three generations of bio-fuels are likely to be developed simultaneously in different countries depending upon local advantages.

However, success of bulk production of bio-fuels on a sustained basis in future, would depend to a great extent on how much we succeed in maintaining climatic conditions congenial for growth of plants and algae used for bio-fuel production. To create favorable climatic conditions, it is imperative that the global warming which has attained an alarming proportion, must be first contained and then reversed at least to the level prevailing 50 years back. This would help restoration of coverage of the glaciers to the original heights and thereby sustain adequate water flow-rate through the rivers which in turn would replenish the ground water reserve and ensure supply of adequate quantity of water for cultivation. Simultaneously, strategic afforestation work should be taken up to stop soil erosion, preferably by planting trees which would supply feed-stocks for bio-fuel production.

Fortunately, the process of bio-fuels production itself should help to control global warming by reducing generation of green house gases; and thereby sustain growth of renewable resources for bio-fuel production. Thus the process of bio-fuel production and controlling global warming are mutually complementary and conform to the basic principle of sustainable growth.

TABLE - 1

RELATIVE OIL-YIELD FROM VARIOUS RENEWABLE RESOURCES

PLANT SPECIES	OIL-YIELD, Gallons/year/acre
CORN	15 – 18 (Alcohol)
SUGARCANE	20 – 25 (Alcohol)
SOYBEAN OIL	45 – 50 (Diesel)
SUNFLOWER OIL	100 – 105 (Diesel)
PALM OIL	600 - 650 (Diesel)
COCONUT OIL	280 – 300 (Diesel)
CASTOR OIL	150 – 155 (Diesel)
OLIVE OIL	120 – 130 (Diesel)
RAPE SEED OIL	125 – 130 (Diesel)
MICRO-ALGAE	1850 – 5000 (Alcohol/ Diesel) (Per acre of pond}

(62a) Greenhouse Gas Legislation, Regulation and Litigation

Mary Ellen Ternes, McAfee & Taft, Oklahoma City, OK

David M. Winfrey, McAfee & Taft, Missouri City, TX

With this presentation, an expert legal team will summarize the latest developments from the climate change summit in Copenhagen, or COP15, implications for U.S. climate change legislation and include an update on status of Senate debate regarding proposed U.S. climate change legislation. The team will also summarize the latest EPA Clean Air Act rulemakings regarding greenhouse gases, including EPA's Mandatory Greenhouse Gas Reporting Rule, Endangerment and Cause or Contribute Finding and new mobile source rules. The team further will review the Clean Air Act Prevention of Significant Deterioration Program and Title V provisions automatically triggered by EPA's mobile source rules, in order to set the foundation for a review of EPA's proposed "Tailoring Rule" to revise major source PSD and Title thresholds for greenhouse gas emitters. Finally, the team will review the latest litigation developments involving major greenhouse gas emission sources.

See more of this Session: Climate Change Legislation and Regulation
See more of this Group/Topical: Environmental Division

IP Strategy Options for Emerging Technologies in the Low-Carbon Energy Space

Abstract: 300-1,500

Paper: 5,000

Authors: Ilian Iliev, CEO, CambridgeIP
ilian.iliev@cambridgeip.com
+44-77-863-73965
Cambridge Intellectual Property Ltd
Sheraton House
Castle Park
Cambridge
CB3 0AX
United Kingdom

Mark Meyer, Business Development Manager-North America, CambridgeIP
mark.meyer@cambridgeip.com
+1-405-312-1412
Cambridge Intellectual Property Ltd
2008 N. Council
Blanchard, Oklahoma 73010
USA

Outline for this Extended Abstract:

1. Importance of Technology (patents), and their role in the growth of low-carbon
    a. Background
    b. Role of IP
2. Market Entry Paths
    a. Cross-Over
    b. Buy Your Way In
    c. Partnerships (Standards)
3. Overview of IP strategy challenges & findings (based on CIP work across 2008-2010)
    a. Solar PV
    b. Biomass & biofuels
    c. Carbon Capture & Storage
4. Summary

Abstract

For technology-based companies patents are a particularly important part of their IP asset base. Private and Public Investment in Low-Carbon Energy Technology continues to grow at a rapid pace as new consortiums, research alliances, venture capital, private equity and technology innovation start-ups are established. Recently the role of Intellectual Property (IP) in Low-Carbon Solutions, commercialization has gained an unusual prominence in the media. Investments by petrochemical companies in novel biofuel processes (backed by strong patent portfolios) have attracted many headlines. Public policy debates in the run-up to the Copenhagen Climate Change summit also saw the role of patents take front stage. But beyond the headlines, how important is IP in the low-carbon energy future? And on a practical level: how can entrepreneurs and company executives maximise the benefit from their patent portfolios and use it to support business objectives? What are the proven IP strategies that can be adopted from other industries to support innovation in the low-carbon energy space?

We provide some insights from patent landscaping and interviews with key industry stakeholders conducted in 2008-9 in clean technology, as well as some suggested implications for Corporate and startup R&D and IP strategies. We begin by overviewing some of the IP strategy challenges already faced by technology companies in the low-carbon energy space: including in solar photovoltaic, biomass, biofuels and carbon capture and storage (CCS). Many of these issues are 'imported' from other industries. For instance in the biofuels space past controversy around standards in fuel production may make the application of future standards for biofuels more difficult to implement; while a migration of litigation trends from the biotech into the biofuels space may also make the scaling up of biofuels production more difficult. We therefore turn next to examples of different strategies pursued by individual companies *and* industries as a whole around how patents are used in the industry's value chain. We examine how technology standards bodies have worked in telecoms and utility metering; and cross-licensing agreements in semi-conductors. We conclude by providing some early examples of emerging standards in the low-carbon energy space, and discuss potential scenarios in areas of interest to the AICHE community.

About the Authors

Ilian Iliev is CEO and co-founder of CambridgeIP and an associate fellow at Chatham House in the UK.

Mark Meyer is Head of Business Development in North America for CambridgeIP.

CambridgeIP is a leading provider of technology business intelligence and IP strategy services to the technology community and policy makers.

IP Role in Technology Growth in Low-Carbon Industries

Background

Intellectual Property (IP) from an awarded patent is a legal document entitling the owner ("assignee") rights to the described property ("claims") for a specific period of time set forth based on where the patent is filed. Properly maintained patents are good for up to 20 years in some jurisdictions, and provide the right to exclude others from manufacturing, selling, importing or offering for sale the protected property covered by the patent claims. To extract value, IP is normally licensed (exclusively or non-exclusively) or may otherwise be sold or transferred.

IP is foundational and provides a platform or spring-board for new ventures and commercialization efforts. Elaborate strategies can be created, based on multiple patents in multiple countries, licensed to specific joint ventures or partners as a path to the market.

Role if IP in New Industry Growth and Maturity

Recent studies (1) have reviewed the corporate business models in various industries and examined how many years they took to evolve; key learnings from this study have been applied to Low Carbon Energy to provide an historical basis for growth projections and possible outcomes. The Low Carbon Energy market-technology space is diverse, but some analogies are thought provoking:

- *Telecommunications industry* – route to maturity required a complex patent technology platform accompanied by capital to build economies of scale and standardization across a global market. Early technology standards were achieved via consortiums and patent pools.

- *Pharmaceutical industry* – trade secrets and early innovation patents provided early growth; but later, cross-licensing across the value chain extended the value to the business of early break-through inventions.

Market Entry Paths

A company can leverage a number of paths to gain market entry. We will examine three specifically. As a framework, the following strategic analysis diagram highlights many of the paths to market entry.

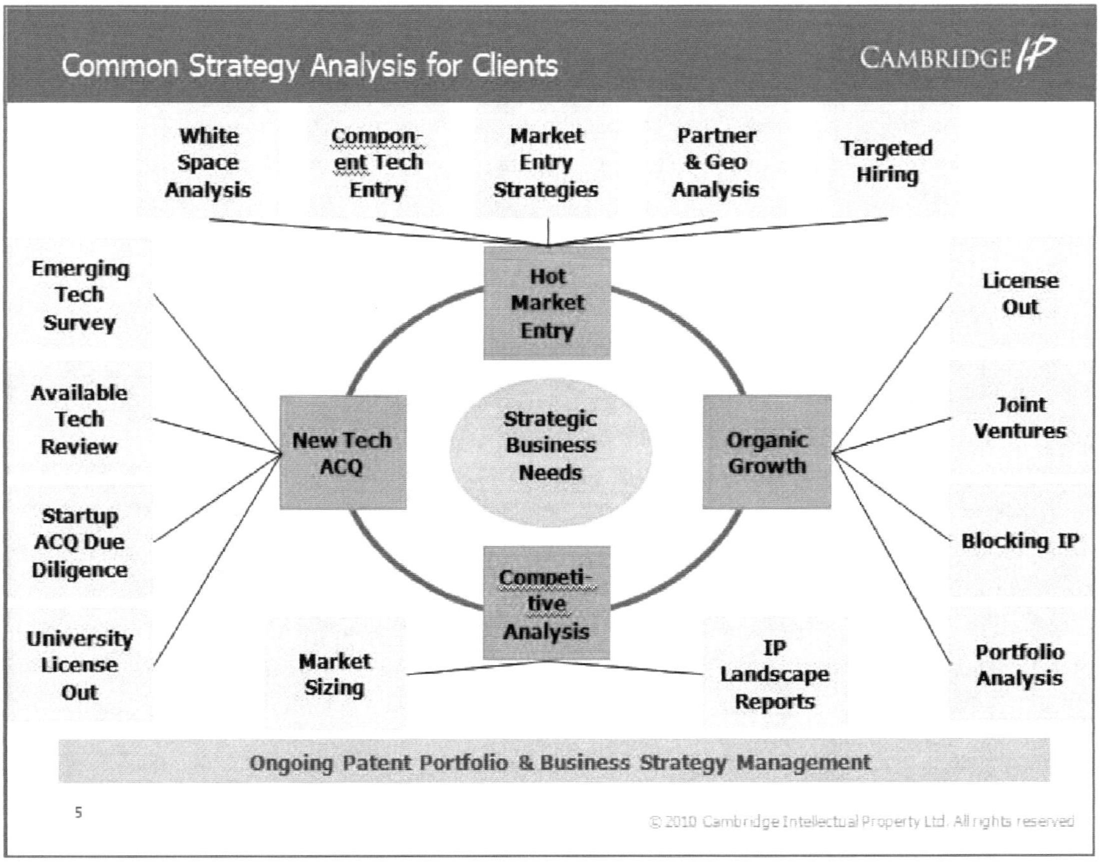

Cross-Over Innovation Path to Growth

The Low Carbon Energy market-technology segment is diverse, drawing technology from "old / mature" sources and many newer technologies. Already, we have seen the potential of technology based in other market-technology sectors to "cross-over" into the emerging Low Carbon market in order to provide the foundation, standardization and early market growth opportunity.

Technology developed and commercialized in other market sectors is often critical to market growth and success in new areas. Examples include clean coal boiler technology in biomass applications, use of rocket engine patents in oxyfuel coal burn and satellite technology uses in solar energy concentrators. Government policies can strongly influence the rates at which technology "cross-over" occurs.

Wind Energy "Cross-Over" Innovation Examples	
Wind Technology Area	***Cross-Over Examples***
Wings; Blades	Aerospace - aerodynamics, wind tunnels Materials - light and strong nanocomposite materials
Offshore Operations	Oil & Gas Offshore Experience - rigging, maintenance, underwater transmission, corrosion management
Software	Meterology - computer modelling and prediction of micro climates

Buying Your Way In

Market entry decisions target and guide initial efforts and strategies in the Low Carbon Energy market. Decisions on focus technologies, freedom to operate, licensing and partnerships establish the strength of your patent portfolio and your access to target buyers. Many larger corporations purchase smaller technology-intensive companies to gain a foot-hold in specific markets. Several transactions in this vein are highlighted below:

Buying Your Way In - Some Recent Examples		
Players	***Deals***	***Markets***
ExxonMobil and Synthetic Genomics	US$600 million	Algae-based Biofuels
BP joint venture with Verenium	US$90 million	Cellulose-based biofuels
GE acquisition of Scanwind	US$19 million	Offshore Wind Turbines
Suzlon purchase of AE-RotorTechniek and Hanson Transmissions	US$338 million IPO	European Wind Market

Partnering for Success (Standards)

Partnerships, Consortiums and related Pilot Projects have been very successful in leveraging collections of IP with Government funding to establish momentum, reduce risk and prove new technology in future markets. Partnerships which aggregate IP (through cross-licensing), in order to create critical mass in a given market, may become more common in Low Carbon Energy markets. To a large extent, innovation centers of excellence provide a perfect venue for collaboration.

Partnerships with successful Pilot efforts are in a stronger position to drive standards in the industry and establish early beach-head positions in the market. Standards provide yet another layer of protection, on top of IP, to protect investments and create competitive barriers to entry.

Overview of IP Strategy Challenges in Low Carbon Energy

CambridgeIP is very active with large corporations, government entities, technology start-ups and universities active in the Low Carbon Energy markets. Highlighted below are some of the key findings of our work in the following areas:

- Solar Photovoltaic (PV)
- Biomass and Biofuels
- Carbon Capture and Storage (CCS)

Solar PV

CambridgeIP has correlated the patent growth with shipments in the Solar PV market. Within the market, we see an overlap of critical technologies which is summarized in the charts below.

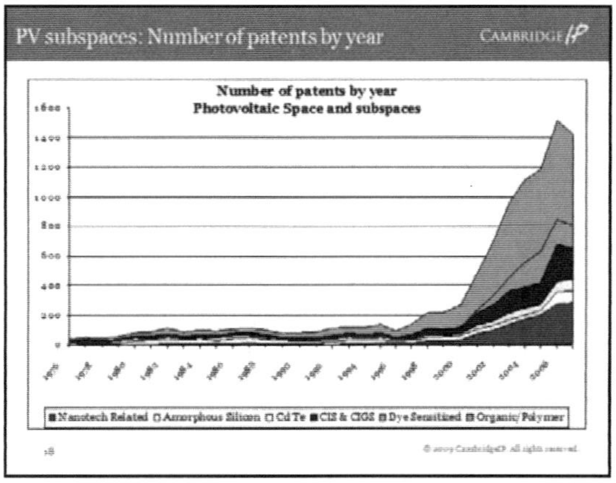

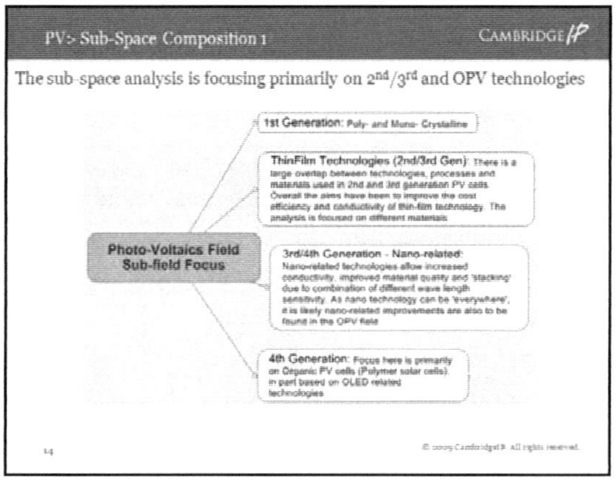

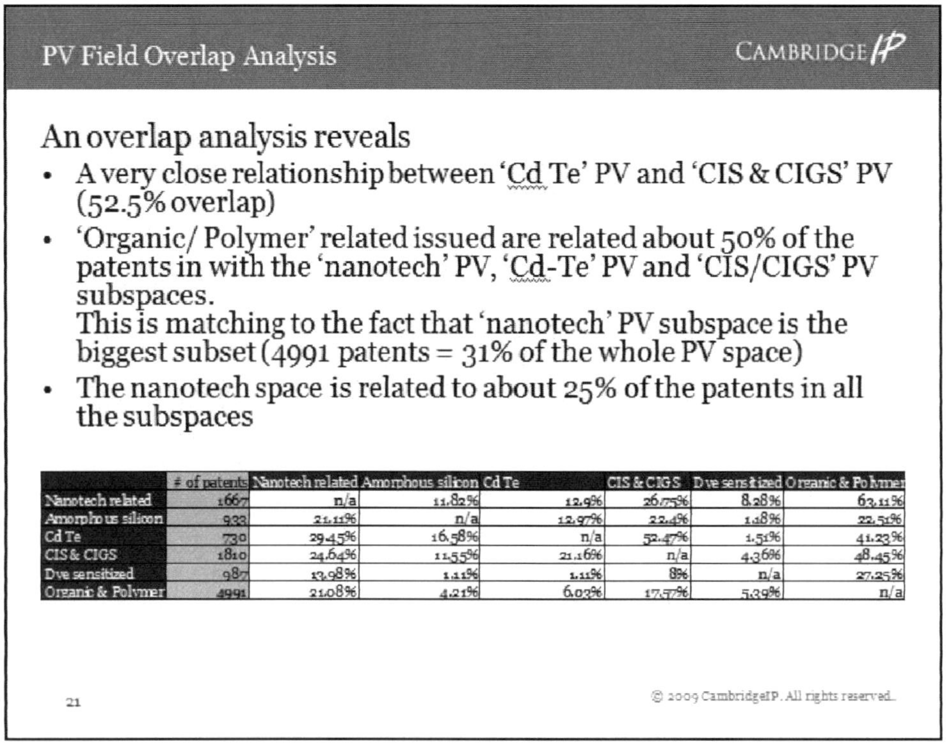

	# of patents	Nanotech related	Amorphous silicon	Cd Te	CIS & CIGS	Dye sensitized	Organic & Polymer
Nanotech related	1667	n/a	11.82%	12.9%	26.75%	8.28%	63.11%
Amorphous silicon	933	21.11%	n/a	12.97%	22.4%	1.18%	22.51%
Cd Te	730	29.45%	16.58%	n/a	52.47%	1.51%	41.23%
CIS & CIGS	1810	24.64%	11.55%	21.16%	n/a	4.36%	48.45%
Dye sensitized	987	13.98%	1.11%	1.11%	8%	n/a	27.25%
Organic & Polymer	4991	21.08%	4.21%	6.03%	17.57%	5.39%	n/a

Key Corporate Motives for Adopting Cross-licensing or patent-pool based technology standards

Industry players that seek to develop and maintain technology leadership in rapidly changing technology spaces can do so by engaging with the wider industry in seeking to accelerate technology diffusion: while staying at the center of the evolving value chain. Our research has identified a number of drivers for other industry players into co-operative IPR arrangements – most of which are of non-monetary, and highly strategic nature.

Table 1. Strategic Drivers for Entering into Co-Operative IPR Arrangements

Considerations	Rationale	Examples
Remain a technology leader	In a rapidly changing industry, remain at the head of technology change	Motorola: semi-conductor cross-licensing
		Nokia: licensing of technology to Siemens
Avoid litigation (defensive and offensive)	Low-cost/reasonable royalties for use of technology: for industry players it is cheaper to license than to risk litigation	Motorola: non-discriminatory/blanket 5-year renewable agreements to both competitors and others
Accelerate innovation	Expose your technology to greater number/type of users	Nokia: licensing of technology to Siemens
Revenue generation	Unilateral licensing out of key IP can generate significant revenues	Motorola: semi-conductor licensing was generating $50mln p.a. in 1990s
Protect value chain against big outsider entrants	Meet challenge to leadership *outside* of industry – retain leadership of the industry	Nokia licensing of S60 platform to counter Windows Mobile entry threat
		Symbian Foundation: royalty-free licensing model to protect against Google Android & Apple
'Increasing the pie'	Change model to redefine the market boundaries & increase services accessible on back of platform	Revenue sources for Smartphones are changing from calls to data and content – even payment services (e.g. Visa mobile payments solutions)

Source: CambridgeIP interviews of industry experts

When evaluating the readiness of an industry for such strategic use of an IP portfolio, companies can conduct an analysis of the industry's readiness for a cross-licensing or technology standards arrangement. Table 2 below illustrates some of these factors.

Table 4: Key Industry Indicators for Industry-readiness for Co-operative IPR

Key Indicators	Industry Examples
Increasing patent complexity & Litigation trends	Increasing number of cross-references between companies' patent portfolios, suggesting complex and overlapping claims of novelty, and the potential for costly patent infringement lawsuits
Increasing speed of product life cycle	R&D investment in vertically integrated systems is too high – need for sharing of IPRs

	Product life cycle becomes shorter than the patent lifespan: sometimes product lines are changed even *before* a patent has been granted (due to patent office examination time lags)
Increased heterogeneity of client/user types	Increasingly complex combinations of product technology requirements, user capabilities and needs, and operator requirements lead to increasing number of 'optimal types' of products that cannot be served by any single company
Fragmentation of market by geography and local conditions	Global spread of a technology standard allows access of large number of market niches; complex combination of service providers/ manufacturers/ users/ value added service providers serving unique and dispersed market niches
Increased complexity of technology platforms	Rapid speed of development and user demand leading to increasing complexity Equipment manufacturing increasingly done by 3rd parties Use of value-added services increasing system complexity
Increasing user autonomy and user innovation	Increasing technology configuration choices, niche types, and combinations between products/services open 'space' for an intermediaries between manufacturers and final users
Major external threat: big player from another industry entering your space	In mobile telephony the entry of Microsoft, Apple and Google in different times of the telecoms industry evolution led to increased technology sharing
Major markets identified, but cannot be reached under current business model	A key driver behind ETSI and GSM standard: to allow *national* EU market leaders to expand internationally into other developed *and developing* markets

Source: Interviews of industry experts

To illustrate this further, we underook an analysis of the Wind Energy sector's readiness for technology standard adoption. A similar analysis can be conducted for cleantech sectors that are closest to the AICHE focus: such as biofuels, carbon capture and Solar PV.

Table 3: Diagnosing the Industry-Readiness for Cross-licensing – Illustrative example for Wind Energy

Key Indicators	Wind Energy Standard Readiness (1-5)
Increasing patent complexity & Litigation trends	**3:** Several high-profile patent litigation cases since 2006: Enercon vs. Vestas; GE vs. Mitsubishi
Increasing speed of product life cycle	**4:** Turbine power has roughly doubled in the last 5 years; rotor diameter in offshore applications may be double the size of onshore
Increased heterogeneity of client/user types	**3:** increasing differences between operator types in terms of geography, purchasing power, micro-wind

	conditions, and off-grid vs. On-grid
Fragmentation of market by geography and local conditions	**4:** highly heterogeneous conditions in terms of regulatory environment; big differences between operational environment in onshore vs. Offshore; but also between micro-wind and mainstream wind
Increased complexity of technology platforms	**4:** increased use of software for optimization, operation in extreme environments, use of composite materials, etc.
Increasing user autonomy (vis-à-vis distributor)	**2**: little user innovation – platforms used as intended; possible move to independence through a retrofit/2nd hand turbines market
Major external threat: big player from another industry entering the space	**2:** many potential entrants, but little chance of paradigm shifters; possibly long-term move to vertical-axis offshore platforms
Major markets identified, but cannot be reached under current business model	**5:** enormous potential for global wind power: it would increase the industry size by multiples; relatively high installation and maintenance costs and lack of storage capacity limit deployment; potential breakthroughs would be through radical de-costing and in power storage

As this illustrative example for wind turbines suggests, the likelihood or need for a cross-licensing agreement, a patent pool, or a standard may vary between the different technology systems. For instance, standardization around communications protocols may ensure that third party software monitoring systems can be implemented across turbines manufactured by other companies, while standardisation of 'plug-in' component of turbines and rotors could allow third party transmission systems to be fit across all turbines. Needless to say, the analysis is **not intended** as a definitive statement, but just as an illustrative example. However, it would be possible to scale up such an approach to develop an analysis across a number of low-carbon energy technology systems and sub-systems, by engaging a broader range of industry stakeholders – including from makers which have not yet been accessed under current business models.

Summary

Patents and strategy around IP portfolios is important for companies seeking to own market share in the Low Carbon Energy markets. IP strength is key to unlocking private and public investment, forming partnerships and creating joint ventures to establish initial market position.

From the analysis conducted by CambridgeIP, we have seen specific market entry strategies be successful, and we have catalogued a number of technology overlap instances which will drive a more complex assessment of the winning strategies for the market-technology mix.

CambridgeIP highlights specific technology areas in Solar PV, Biomass and CCS to draw-out some the strategies and challenges in the Low Carbon Energy markets. We hope that these examples assist in the development of consortiums and emerging standards to propel growth of Low Carbon Energy technology commercialization for the AICHE community.

We also provided an outline of a diagnostic tool that can assist leading industry players to evaluate their industry's readiness for a move to a cross-licensing/patent pool structure that can accelerate technology diffusion: while retaining leadership in a rapidly changing technology value chain.

Citations

(1) December 2009, Renewable Energy Focus, Finance Supplement "The importance of IP", Ilian Iliev and Mark Meyer

(90a) Complying with the EPA Mandatory GHG Reporting Rule, 40 C.F.R. Part 98

Mary Ellen Ternes, McAfee & Taft, Oklahoma City, OK

David M. Winfrey, McAfee & Taft, Missouri City, TX

Glen A. Capra, Weston Solutions, Inc., Austin, TX

This workshop session provides a hands-on review and discussion of EPA's Mandatory Reporting of Greenhouse Gases (GHG) Rule effective December 29, 2009. The workshop will be facilitated by expert environmental legal counsel and technical consultants. The format will include a review of the purpose of the rule, GHG Monitoring Plan requirements, transitioning from Best Available Monitoring Methods to the specific GHG monitoring and emission calculation methods, quality assurance, missing data calculations, record retention, designated representatives and agreements, and enforcement. The review will also work through compliance with Subpart C (stationary fuel combustion) requirements. Questions and discussion will be encouraged throughout the review. With this abstract, the Session Chair, Mary Ellen Ternes, requests that, prior to March 2010, attendees e-mail specific questions or areas of discussion to be included in the review to Maryellen.ternes@mcafeetaft.com identifying the Subject as: "AICHE General GHG Workshop – Question."

See more of this Session: Greenhouse Gas Emission Reporting

See more of this Group/Topical: Environmental Division

(91a) Analysis of Greenhouse Gas (GHG) Regulation Impacts to the U.S. Refining Industry

Ian M. Glasgow, Mustang Engineering, L.P., Houston, TX

Stan Polcar, Mustang Engineering, L.P., Houston, TX

R.E. (Ed) Palmer, Mustang Engineering, L.P., Houston, TX

President Obama has stated that reduction of greenhouse gases (GHG) and reduced dependence to foreign oil is part of his immediate legislative agenda. At the time this abstract is written, the Waxman-Markey bill that will reduce greenhouse gases (GHG) emissions was passed by the House in June 2009; the Senate has passed a similar bill (Kerry-Boxer bill) out of Energy & Environment committee. The EPA has determined that endangerment to human health and welfare exists due to climate change from GHG and is actively pursuing options to utilize the Clean Air Act for GHG emission reductions. Updated CAFÉ standards have been established that will take effect in 2016 to increase light duty vehicle efficiency to 35.5 mpg and include GHG emissions as a part of vehicle standards. Previously passed legislation from the 2007 Energy Independence Act (EIA) will also impact refiners with a mandated 36 billion gallons per year of biofuel produced by 2022 under the Renewable Fuel Standard (RFS-2).

In this paper various legislation on GHG and energy are reviewed, with the focus being on the proposed Waxman-Markey American Clean Energy and Security Act and the EPA currently proposed RFS-2 that is a part of the 2007 EIA.

A brief overview of Life Cycle Assessment (LCA) and cap-and-trade systems are presented for information purposes. Case studies for a notional 150 MBPSD refinery are analyzed to evaluate the trends and impacts to a refinery resulting from a cap-and-trade system (Waxman-Markey) where product emissions are included. Comparison of the Waxman-Markey bill to the increased CAFÉ and RFS-2 regulations is also provided. The major GHG manufacturing emissions sources of a notional refinery are identified and a preliminary discussion of potential solutions to reduce emissions including the use of petroleum coke gasification for CCS and offsetting hydrogen plant CO2 emissions is highlighted.

See more of this Session: Greenhouse Gases Sequestration
See more of this Group/Topical: Topical 6: 10th Topical Conference on Gas Utilization

(91b) High Capacity Regenerable Sorbent for Pre-Combustion CO_2 Capture

Gökhan Alptekin, TDA Research Inc., Wheat Ridge, CO

Ambalavanan Jayaraman, TDA Research Inc., Wheat Ridge, CO

Steve Dietz, TDA Research Inc., Wheat Ridge, CO

Margarita Dubivik, TDA Research Inc., Wheat Ridge, CO

Lauren Brickner, TDA Research Inc., Wheat Ridge, CO

TDA is developing a novel sorbent that removes CO_2 via physical adsorption from both synthesis gas, while the relatively strong affinity of the sorbent for CO_2 enables effective operation at temperatures up to 300°C (well above the dew point of synthesis gas stream generated by most commercial gasifiers). However, because the sorbent and the CO_2 do not form a true covalent bond, the energy needed to regenerate our sorbent (5.4 kcal per mol of CO_2) is much lower than that observed for either chemical absorbents (e.g., 29.9 kcal/mol CO_2 for sodium carbonate) or amine-based solvents (e.g., 14.2 kcal/mol CO_2 for monoethanolamine). Our sorbent can be regenerated isothermally and CO_2 can be recovered at pressure (~150 psia). Thus, the energy needed to regenerate the sorbent and compress the CO_2 for sequestration is significantly lower than that for any technology reported to date. TDA's CO_2 sorbent could be used to capture CO_2 in a Integrated Gasification Combined Cycle (IGCC) power plant after the water-gas-shift (WGS) reaction when most of the carbon is in the form of CO_2. The high surface area and favorable porosity of the sorbent also provide a unique platform to introduce additional functionality, such as active groups to catalyze the or to remove trace metals (e.g., Hg, As).

We carried out bench-scale evaluations of TDA's CO_2 sorbent in both fixed-bed adsorber and a thermo gravimetric analyzer (TGA). The sorbent achieved high CO_2 capacity, high removal efficiency and excellent durability. The sorbent showed stable performance in both these systems under simulated synthesis gas (44 cycles in the bench-scale flow system and over 1,644 cycles in the TGA). We also studied the effect of operating parameters like temperature, pressure, and cycle time on the sorbent performance. The results from our on-going sorbent evaluations will be presented at the meeting.

See more of this Session: Greenhouse Gases Sequestration
See more of this Group/Topical: Topical 6: 10th Topical Conference on Gas Utilization

(91c) The Third C in CCS - Compression

Dipanjan Bhattacharya, Bechtel Oil and Gas, Houston, TX

Ramachandra P. Tekumalla, Advanced Simulation & Analysis, Bechtel Corporation, Houston, TX

David Messersmith, Bechtel, Houston, TX

Lee Schmoe, Bechtel Corporation, Houston, TX

Tomas Maramba, Bechtel, Houston, TX

The capture and sequestration of carbon dioxide emitted from process plants is one of the more pressing issues for the energy industry. While much has been reported on the capture of CO2, little has been published on the compression options and issues. This paper summarizes the results of a Bechtel investigation for three different plant applications: a pulverized coal-fired power plant, an LNG plant processing high CO2 natural gas, and an IGCC complex. Working closely with equipment suppliers, Bechtel examined capital and operating costs for several configurations including a series of compressors, a combination of compressor and pumps (supercritical and refrigerated liquids), and different dehydration designs.

The paper provides a framework for the readers to more-effectively initiate and complete their own project-specific designs.

See more of this Session: Greenhouse Gases Sequestration
See more of this Group/Topical: Topical 6: 10th Topical Conference on Gas Utilization

(114a) Catalytic Dehydrogenation of Methylcyclohexane for the On-Board Hydrogen Storage and Supply

Muhammad R. Usman, School of Chemical Engineering and Analytical Science, The University of Manchester, Manchester, United Kingdom

David L. Cresswell, School of Chemical Engineering and Analytical Science, The University of Manchester, Manchester, United Kingdom

Arthur A. Garforth, School of Chemical Engineering and Analytical Science, The University of Manchester, Manchester, United Kingdom

In the current scenario of the energy demands, nations require alternative energy supplies that are not only sustainable and economical, but benign to the environment. A hydrogen based energy system is considered one among the possible alternatives. Cheap and safe storage of hydrogen is probably the greatest hurdle to the success of the hydrogen economy. The Methylcyclohexane-Toluene-Hydrogen (MTH) system is a safe and economical way for the storage and on-board hydrogen generation. Also, the existing motor fuel infrastructure can effectively be utilized for the transportation and dispensation of liquid methylcyclohexane. Commercial catalysts as well as in-house were studied for the range of reaction conditions to screen the best catalytic material required for the dehydrogenation reaction. The in-house catalyst was found not only highly active, selective towards toluene, but also with acceptable life. Rigorous kinetic analyses of the data were carried out for elucidating the reaction mechanism and for the development of an appropriate kinetic model. Based on the kinetic equation developed above, mathematical modeling and simulation of the dehydrogenating reactor was carried out for the optimum performance of the on-board dehydrogenating reactor.

See more of this Session: H2/CO2 Generation/Separation/Storage/Utilization

See more of this Group/Topical: Environmental Division

(114b) Molecular Simulations and Predictions of Pure and Mixed CO_2/H_2 Gas Absorption Into Some Ionic Liquids

Wei Shi, Computational Chemistry, National Energy Technology Laboratory, Parsons Corporation, Pittsburgh, PA

Dan C. Sorescu, Computational Chemistry, National Energy Technology Laboratory, Pittsburgh, PA

Classical molecular dynamics and Monte Carlo simulations are used to calculate the self-diffusivity and solubility of pure and mixed gases of CO_2, H_2, and Ar absorbed into some ionic liquids. For H_2 absorption into [hmim][Tf2N], the computed isotherms, Henry's law constants, and partial molar enthalpies agree very well with the experimental data obtained by Maurer et al. [J. Chem. Eng. Data 2006, 51, 1364], however they disagree with those by Noble et al. [Ind. Eng. Chem. Res. 2008, 47, 3453] and Costa Gomes [J. Chem. Eng. Data 2007, 52, 472]. Simulations for H_2 absorption into another ionic liquid of [bmim][PF6] also agree with the experimental data obtained by Maurer et al. The interaction between H_2 and [hmim][Tf2N] is very weak, about three times smaller than Ar, and six times smaller than CO_2. These results are consistent with a decreasing order in the solubility from CO_2 to Ar and to H_2. The molar volume of the ionic liquid is found to be the determining factor for the H_2 solubility. The simulated permeability values for CO_2 and H_2 in [hmim][Tf2N] are clsoe to the experimental data by Luebke et al. [J. Membr. Sci. 2007, 2008].

Based on our simulations, we propose to seacrh for ionic liquids with small molar volume leading to low H_2 solubility, but still having high CO_2 solubility through non-physical interaction for the CO_2/H_2 separation. We have identified several ionic liquids and they are predicted to show high CO_2/H_2 selectivity.

See more of this Session: H2/CO2 Generation/Separation/Storage/Utilization
See more of this Group/Topical: Environmental Division

(114c) Elastic Layered Metal-Organic Framework Absorbents for Carbon Dioxide Capture

Christian M. Lastoskie, Civil & Environmental Engineering, University of Michigan, Ann Arbor, MI

Tran D. Trinh, Civil & Environmental Engineering, University of Michigan, Ann Arbor, MI

Elastic layered metal-organic framework (ELM) adsorbents exhibit a flexible two-dimensional latent porous crystalline structure. ELMs undergo abrupt reversible gated sorption transitions from an empty collapsed structure to a filled expanded porous state through cooperative adsorption of guest molecules between layer planes. The ELM gating transition and associated hysteresis loop is different than the gas condensation phase transition commonly observed for other nanoporous adsorbents, in that it is induced by rearrangement of the MOF structure rather than by sorbent pore size distribution effects. Certain ELMs have very high selectivity for CO2 adsorption from gas mixtures, and are intriguing adsorbents for cost reductions in carbon capture and storage operations for CO2 removal from nitrogen and other flue gases in post-combustion separations or from hydrogen in pre-combustion separation (e.g. following coal gasification). Moreover, some ELMs have attractive sorption capacities, enthalpies, and isotherm features for efficient recovery of captured CO2 using temperature or pressure swing adsorption. ELMs thus merit further investigation as materials for sorbent-based capture of industrial CO2 emissions. We report results from molecular simulations of ELM-11 [Cu(BF4)2(bpy)2 (bpy = bipyridine)] and ELM-12 [Cu(bpy)2(OTf)2 (OTf = trifluoromethanesulfonate)] for separation of CO2 from pre- and postcombustion gas mixtures. Adsorption capacity, CO2 selectivity, and isosteric heat of adsorption for pure CO2 and binary CO2/H2 and CO2/N2 mixtures were calculated via grand canonical Monte Carlo (GCMC) simulation at temperatures (273 to 400 K), pressures, and compositions (5 to 15% CO2) representative of power plant flue gas streams. GCMC simulations of CO2 capacity above the gate pressure using a rigid model of ELM-11 agreed well with experimental isotherms at 273 K. CO2 sorption capacity (140 mg/g) and heat of adsorption on ELM-11 are competitive with commodity adsorbents at realistic process conditions. Simulations project a CO2/N2 selectivity of 700 at 300 K (Figure 1), a capture ratio significantly higher than common activated carbons and zeolites. The simulated CO2 selectivity is comparable to experimental results for separating CO2 from N2 and O2 by temperature swing adsorption from 268 K to 311 K using ELM-11. Desorption of captured CO2 of more than 99% purity was obtained from a ternary mixture of 40% CO2, 48% N2 and 12% O2 at 268 K. GCMC simulations on ELM-11 for CO2/H2 mixtures representing shifted synthesis gas were also carried out and CO2 selectivities of 1000 and 400 were obtained at 298 and 373 K respectively. These compare favorably with reported values for commercial adsorbents used for gas separations. To investigate framework flexibility effects on CO2 sorption in ELM-11, a hybrid Monte Carlo/molecular dynamics (MCMD) method was applied. CO2 adsorption isotherms showed good agreement with experiments at 273 K. Mixed-gas simulations yield CO2/N2 selectivities of 70 and 20 respectively at 300 K and 350 K, comparable or superior to other common adsorbents, but substantially lower than selectivities predicted from rigid model framework simulations. Isotherms for CO_2 and N_2 on ELM-12 are contrasted to those obtained for ELM-11. The bulkier ELM-12 counter-ion yields a structure with microporosity in addition to latent porosity as found in ELM-11. Unlike ELM-11, ELM-12 exhibits a two-step isotherm with micropore filling followed by interlayer clathrate formation at a narrowly defined gating pressure.

See more of this Session: H2/CO2 Generation/Separation/Storage/Utilization
See more of this Group/Topical: Environmental Division

A Solid-Gas Ideal Gibbs Reactor Model for Natural Gas Reforming in the Presence of CaO

Jorge Pena Lopez[1], *Vasilios I. Manousiouthakis*[2],
Chemical and Biomolecular Engineering Department,
Hydrogen Engineering Research Consortium,
University of California, Los Angeles

Introduction

An ongoing increase of carbon dioxide concentrations in the atmosphere during the last few decades has been well documented. This has led to concerns regarding the possible consequences of this phenomenon, resulting in tighter industrial regulation of carbon dioxide emissions. Hydrogen has been proposed as a viable and effective transportation energy carrier for small and medium vehicles. Natural gas reforming is the main industrial route for hydrogen production. The process consists of a first step in which a gaseous mixture of natural gas (methane) and water reacts at high temperature (about 1000 K) and medium pressure (5-20 bar) to generate a mixture of methane, water, carbon dioxide, carbon monoxide and hydrogen. This mixture is then reacted at lower temperatures (475 K -600 K), in a second step occurring in a gas-shift reactor sequence that aims to eliminate the carbon monoxide from the mixture. The overall process can be described by the following equation:

$$CH_4^{(g)} + 2H_2O^{(g)} \rightleftarrows CO_2^{(g)} + 4H_2^{(g)} \qquad (1)$$

This suggests the equilibrium can be shifted towards hydrogen production by depletion of carbon dioxide. The addition of calcium oxide to the reforming system has the potential

[1] Email address: jpenalopez@ucla.edu
[2] To whom correspondence should be addressed. Email address: vasilios@ucla.edu

to alter the equilibrium and at the appropriate conditions lead to the generation of calcium carbonate, consuming the carbon dioxide generated by the forward direction in equation 1 and shifting the equilibrium towards the reforming products.

$$CaO + CO_2 \rightleftarrows CaCO_3 \tag{2}$$

A potential complication stems from the fact that calcium oxide could react with water to produce calcium hydroxide thus preventing the calcium oxide from removing the carbon dioxide and at the same time reducing the amount of water available for reforming.

$$CaO + H_2O \rightleftarrows Ca(OH)_2 \tag{3}$$

To quantify this effect, we carry out equilibrium calculations minimizing the overall Gibbs free energy for the ideal gas-solid system: $CH_4^{(g)}$, $CO_2^{(g)}$, $CO^{(g)}$, $H_2O^{(g)}$, $H_2^{(g)}$, $CaO^{(s)}$, $CaCO_3^{(s)}$, $Ca(OH)_2^{(s)}$, $C^{(s)}$. The optimization problem is formulated as follows:

$$\underset{N_i^{(j)}, \mu_i^{(j)}}{\text{minimize}} \sum_{i=1}^{NC} \sum_{j=1}^{NP} N_i^{(j)} \mu_i^{(j)} \tag{4}$$

Subject to:

$$\mathbf{A} \cdot [\sum_{j=1}^{NP} \mathbf{N}^{(j)} - \mathbf{N}^*] = 0 \tag{5}$$

$$\mu_i^{(1)} = \mu_i^{(j)} = \Delta G_{if}^{(j)}(T) + RT \ln\left(\frac{\hat{f}_i^{(j)}}{f_i^{o(j)}}\right); \; i = 1, NC; \; j = 1, NP \tag{6}$$

$$N_i^{(j)} \geq 0; \; i = 1, NC; \; j = 1, NP \tag{7}$$

where all symbols are defined in the notation section.

At fixed temperature, the multiphase total Gibbs free energy minimization problem for the ideal Gibbs gas-solid reactor considered here is non-linear and convex and can thus be solved globally using a NLP optimization solver (MINOS 5.51 was used). In turn this implies that equilibrium calculations can rapidly and correctly be carried out for various inlet , temperature and pressure conditions.

Results and Discussion

The results of our study identify equilibrium compositions for all species as a function of temperature and pressure and establish the ranges within which this process can by implemented industrially. The effect of the presence of calcium oxide in the system is positive within certain temperature and pressure ranges and can be summarized as follows:

1. High hydrogen yield attained at temperatures lower than those of conventional reforming.
2. Solid carbon formation greatly reduced.
3. Carbon dioxide and monoxide concentrations near zero.
4. Higher hydrogen concentration.

These indicate the potential to carry out reforming at lower temperatures and pressures. Conventional reforming is carried out at high pressures to reduce the formation of carbon onto the catalyst. However, hydrogen production is clearly favored at low pressures according to the Le Chatelier principle. Addition of CaO to the system reduces carbon formation allowing the process to be carried out at lower pressures than conventional reforming. Negligible carbon dioxide concentrations are the result of calcium carbonate formation; furthermore, carbon monoxide is just as scarce as carbon dioxide thus possibly negating the need for a shift reactor sequence to eliminate monoxide. By reducing the quantities of other gaseous components, the concentration of hydrogen is greatly increased thus reducing the cost of its purification process.

Appendix A

Proof of convexity for ideal Gibbs free energy objective function

Let $G\left(\left\{N_i^{(j)}\right\}_{\substack{i=1,\dots,NC\\ j=1,\dots,NP}}\right)=\sum_{i=1}^{NC}\sum_{j=1}^{NP}N_i^{(j)}\mu_i^{(j)}$

be the general expression for the total Gibbs free energy of a system with NC components and NP phases. Let system Ω contain n components, and $s+l+1$ phases; where s phases are solid, l are liquid, and 1 is gaseous. This expression can be rewritten as follows:

$$\underbrace{\sum_{j=1}^{s}\sum_{i=1}^{n}N_i^{(j)}\mu_i^{(j)}}_{solid}+\underbrace{\sum_{j=s+1}^{s+l}\sum_{i=1}^{n}N_i^{(j)}\mu_i^{(j)}}_{liquid}+\underbrace{\sum_{i=1}^{n}N_i^{(s+l+1)}\mu_i^{(s+l+1)}}_{gas}$$

We know that:

$$\mu_i^{(j)}=\mu_{if}^{(j)}+RT\ln\left(\frac{\hat{f}_i^{(j)}}{f_i^{o(j)}}\right)$$

and,

$$\frac{\hat{f}_i^{(j)}}{f_i^{o(j)}}=\begin{cases}1 & \text{if } (j) \text{ solid phase}\\ x_i^{(j)} & \text{if } (j) \text{ ideal liquid solution, } x_i^{(j)}=\dfrac{N_i^{(j)}}{N^{(j)}}\\ y_i^{(j)}\left(\dfrac{P}{P^o}\right) & \text{if } (j) \text{ ideal gas mixture, } y_i^{(j)}=\dfrac{N_i^{(j)}}{N^{(j)}}\end{cases}$$

The total Gibbs free energy for the system Ω can then be expressed as:

$$G\left(\left\{N_i^{(j)}\right\}_{\substack{i=1,\ldots,NC\\ j=1,\ldots,NP}}\right)=\sum_{j=1}^{s}\sum_{i=1}^{n}N_i^{(j)}\left(\mu_{if}^{\ (j)}+RT\ln\left(\frac{\hat{f}_i^{(j)}}{f_i^{o(j)}}\right)\right)$$

$$+\sum_{j=s+1}^{s+l}\sum_{i=1}^{n}N_i^{(j)}\left(\mu_{if}^{\ (j)}+RT\ln\left(\frac{\hat{f}_i^{(j)}}{f_i^{o(j)}}\right)\right)$$

$$+\sum_{i=1}^{n}N_i^{(s+l+1)}\left(\mu_{if}^{\ (s+l+1)}+RT\ln\left(\frac{\hat{f}_i^{(s+l+1)}}{f_i^{o(s+l+1)}}\right)\right)$$

where $\mu_{if}^{\ (j)}$ are constants corresponding to the Gibbs free energy of formation.

<u>We show that $G\left(\left\{N_i^{(j)}\right\}_{\substack{i=1,\ldots,NC\\ j=1,\ldots,NP}}\right)$ is convex over $N_i^{(j)}$</u>. Rewriting $G\left(\left\{N_i^{(j)}\right\}_{\substack{i=1,\ldots,NC\\ j=1,\ldots,NP}}\right)$ to group the linear and nonlinear terms we obtain:

$$G\left(\left\{N_i^{(j)}\right\}_{\substack{i=1,\ldots,NC\\ j=1,\ldots,NP}}\right)=G\left(\left\{N_i^{(j)}\right\}_{\substack{i=1,\ldots,NC\\ j=1,\ldots,NP}}\right)\Bigg|_{lin}+G\left(\left\{N_i^{(j)}\right\}_{\substack{i=1,\ldots,NC\\ j=1,\ldots,NP}}\right)\Bigg|_{nonlin}$$

$$G\left(\left\{N_i^{(j)}\right\}_{\substack{i=1,\ldots,NC\\ j=1,\ldots,NP}}\right)\Bigg|_{lin}=\sum_{j=1}^{s}\sum_{i=1}^{n}N_i^{(j)}\mu_{if}^{\ (j)}+\sum_{j=s+1}^{s+l}\sum_{i=1}^{n}N_i^{(j)}\mu_{if}^{\ (j)}$$

$$+\sum_{i=1}^{n}N_i^{(s+l+1)}\left(\mu_{if}^{\ (s+l+1)}+RT\ln\left(\frac{P}{P^o}\right)\right)$$

$$G\left(\left\{N_i^{(j)}\right\}_{\substack{i=1,\ldots,NC\\ j=1,\ldots,NP}}\right)\Bigg|_{nonlin}=\sum_{j=1}^{s}\sum_{i=1}^{n}N_i^{(l)}\left(RT\ln\left(\frac{N_i^{(l)}}{\sum_{i=1}^{n}N_i^{(l)}}\right)\right)+\sum_{j=s+1}^{s+l}\sum_{i=1}^{n}N_i^{(g)}\left(RT\ln\left(\frac{N_i^{(g)}}{\sum_{i=1}^{n}N^{(g)}}\right)\right)$$

$$\frac{G\left(\left\{N_i^{(j)}\right\}_{\substack{i=1,\ldots,NC\\ j=1,\ldots,NP}}\right)\Bigg|_{nonlin}}{RT}=\sum_{j=1}^{s}\sum_{i=1}^{n}N_i^{(l)}\left(\ln\left(\frac{N_i^{(l)}}{\sum_{i=1}^{n}N_i^{(l)}}\right)\right)+\sum_{j=s+1}^{l}\sum_{i=1}^{n}N_i^{(g)}\left(\ln\left(\frac{N_i^{(g)}}{\sum_{i=1}^{n}N^{(g)}}\right)\right)$$

Let $F:\mathbb{R}^n\to\mathbb{R}$ with domain in the open $\mathbb{R}^n_{++}$ space be defined as:

$$\mathrm{F}(\mathbf{N})=\sum_{i=1}^{n}N_i\ln\left(\frac{N_i}{\sum_{k=1}^{n}N_k}\right),\quad \mathbf{N}=\begin{bmatrix}N_1 & \cdots & N_n\end{bmatrix}^{\mathrm{T}}\in\mathbf{dom}\,\mathrm{F}$$

F(**N**) is a convex function.

Proof:

$$\frac{\partial\,\mathrm{F}(\mathbf{N})}{\partial N_j}=\frac{\partial\sum_{i=1}^{n}N_i\ln\left(\frac{N_i}{\sum_{k=1}^{n}N_k}\right)}{\partial N_j}=\sum_{i=1}^{n}\frac{\partial\left(N_i\ln\frac{N_i}{\sum_{k=1}^{n}N_k}\right)}{\partial N_j},\quad j=1,\ldots,n\Rightarrow.$$

$$\frac{\partial\,\mathrm{F}(\mathbf{N})}{\partial N_j}=\ln\frac{N_j}{\sum_{k=1}^{n}N_k},\quad j=1,\ldots,n\Rightarrow$$

$$\nabla^2\,\mathrm{F}(\mathbf{N})=\left\{\begin{array}{l}\dfrac{\partial^2\,\mathrm{F}(\mathbf{N})}{\partial N_j^{\,2}}=\dfrac{1}{N_j}-\dfrac{1}{\sum_{k=1}^{n}N_k}\\[2ex] \underset{j\neq i}{\dfrac{\partial^2\,\mathrm{F}(\mathbf{N})}{\partial N_j\partial N_i}}=\underset{j\neq i}{\dfrac{\partial}{\partial N_i}}\left(\ln\left(N_j\right)-\ln\left(\sum_{k=1}^{n}N_k\right)\right)=-\dfrac{1}{\sum_{k=1}^{n}N_k}\end{array}\right\}\Rightarrow$$

$$\nabla^2\,\mathrm{F}(\mathbf{N})=\begin{bmatrix}\frac{1}{N_1}-\frac{1}{\sum_{k=1}^{n}N_k} & \frac{-1}{\sum_{k=1}^{n}N_k} & \cdots & \frac{-1}{\sum_{k=1}^{n}N_k}\\ \frac{-1}{\sum_{k=1}^{n}N_k} & \frac{1}{N_2}-\frac{1}{\sum_{k=1}^{n}N_k} & & \vdots\\ \vdots & & \ddots & \\ \frac{-1}{\sum_{k=1}^{n}N_k} & \cdots & & \frac{1}{N_n}-\frac{1}{\sum_{k=1}^{n}N_k}\end{bmatrix}$$

$\nabla^2\,\mathrm{F}(\mathbf{N})\geq 0$ (positive semidefinite) if and only if $\mathrm{F}(\mathbf{N})$ is convex.

We know that $\nabla^2\,\mathrm{F}(\mathbf{N})\geq 0$ if and only if the determinants of all leading principal submatrices of $\nabla^2\,\mathrm{F}(\mathbf{N})$ are nonnegative. Let $\left|\nabla^2\,\mathrm{F}(\mathbf{N})\right|_{m\times m}$ be the m-th sized leading principal submatrix of $\nabla^2\,\mathrm{F}(\mathbf{N})$. Then,

$$\left|\nabla^2 \mathrm{F}(\mathbf{N})\right|_{m\times m} = \begin{bmatrix} \frac{1}{N_1} - \frac{1}{\sum_{k=1}^{n} N_k} & \frac{-1}{\sum_{k=1}^{n} N_k} & \cdots & \frac{-1}{\sum_{k=1}^{n} N_k} \\ \frac{-1}{\sum_{k=1}^{n} N_k} & \frac{1}{N_2} - \frac{1}{\sum_{k=1}^{n} N_k} & & \vdots \\ \vdots & & \ddots & \\ \frac{-1}{\sum_{k=1}^{n} N_k} & \cdots & & \frac{1}{N_m} - \frac{1}{\sum_{k}^{n} N_k} \end{bmatrix}$$

This matrix can be rewritten as:

$$\left|\nabla^2 \mathrm{F}(\mathbf{N})\right|_{m\times m} = \begin{bmatrix} \frac{1}{N_1} & 0 & \cdots & 0 \\ 0 & \frac{1}{N_2} & & \vdots \\ \vdots & & \ddots & \\ 0 & \cdots & & \frac{1}{N_m} \end{bmatrix} - \frac{1}{\sum_{k=1}^{n} N_k} \begin{bmatrix} 1 & 1 & \cdots & 1 \\ 1 & 1 & & \\ \vdots & & \ddots & \\ 1 & & & 1 \end{bmatrix} \Rightarrow$$

$$\left|\nabla^2 \mathrm{F}(\mathbf{N})\right|_{m\times m} = \begin{bmatrix} \frac{1}{N_1} & 0 & \cdots & 0 \\ 0 & \frac{1}{N_2} & & \vdots \\ \vdots & & \ddots & \\ 0 & \cdots & & \frac{1}{N_m} \end{bmatrix} - \frac{1}{\sum_{k=1}^{n} N_k} \begin{bmatrix} 1 \\ 1 \\ \vdots \\ 1 \end{bmatrix} \begin{bmatrix} 1 \\ 1 \\ \vdots \\ 1 \end{bmatrix}^{\mathrm{T}} \Rightarrow$$

Let $\mathbf{A_m} = \begin{bmatrix} \frac{1}{N_1} & 0 & \cdots & 0 \\ 0 & \frac{1}{N_2} & & \vdots \\ \vdots & & \ddots & \\ 0 & \cdots & & \frac{1}{N_m} \end{bmatrix}$, $\mathbf{u_m} = -\frac{1}{\sum_{k=1}^{n} N_k} \begin{bmatrix} 1 \\ 1 \\ \vdots \\ 1 \end{bmatrix}$, $\mathbf{v_m} = \begin{bmatrix} 1 \\ 1 \\ \vdots \\ 1 \end{bmatrix}$. Then $\nabla^2 \mathrm{F}(\mathbf{N}) = \mathbf{A} + \mathbf{u}_m \mathbf{v}_m^{\mathbf{T}}$

From Sylvester's determinant lemma $\det\left(\mathbf{A}+\mathbf{u}_m\mathbf{v}_m^{\mathbf{T}}\right)=\left(1+\mathbf{v}_m^{\mathbf{T}}\mathbf{A}^{-1}\mathbf{u}_m\right)\det\left(\mathbf{A}\right)$.

Then, $\det\left(\left|\nabla^2 \mathrm{F}(\mathbf{N})\right|_{m\times m}\right)$ becomes:

$$\det\left(\left|\nabla^2 \mathrm{F}(\mathbf{N})\right|_{m\times m}\right)=\frac{1}{\prod_{i=1}^{m} N_i}\left(1-\frac{\sum_{k=1}^{m} N_k}{\sum_{k=1}^{n} N_k}\right)\geq 0,\ \forall m\in\{1,\ldots,n\}$$

The above expression shows that F(**N**) is a convex function since its Hessian is positive semidefinite (the determinants of all leading principal submatrices of F(N) are nonnegative). Since $G\left(\left\{N_i^{(j)}\right\}_{\substack{i=1,\ldots,NC\\ j=1,\ldots,NP}}\right)$ is a positive linear combination of convex functions and affine functions, then it is convex□.

Notation:

NC	number of components.
NP	number of phases.
$N_i^{(j)}$	number of moles of component *i* in the *j*-th phase.
$\mathbf{A}$	atomic balance matrix.
$\mathbf{N}^*$	initial number of moles vector.
$\mu_i^{(j)}$	chemical potential of component *i* in the *j*-th phase.
$\Delta G_{if}^{(j)}(T)$	ideal Gibbs free energy of formation for component *i*, in phase *j* at T
$\hat{f}_i^{(j)}, f_i^{o(j)}$	fugacity of component *i* in the *j*-th phase, fugacity of component *i* in the *j*-th phase at standard conditions.

Acknowledgements

This work was funded partially by UC Mexus CONACyT and National Science Foundation NSF-CBET 0829211 to which we are gratefully thankful.

Hydrogen Fill-Up Pressure and Temperature Modeling

F. Olmos, S. Munoz, C. Garcia, A. Garcia, M. Pantoja, J. Garrett, V. Manousiouthakis*

Department of Chemical and Biomolecular Engineering, Hydrogen Engineering Research Consortium (HERC),

University of California at Los Angeles (UCLA), CA 90095-1592, United States

*To whom all correspondence should be addressed at vasilios@ucla.edu

Introduction

One of the challenges in the deployment of hydrogen vehicles is their rapid fuel replenishment. If the flowrate of hydrogen fed to the vehicle is low, then the fill-up time is excessive, while if the flowrate is high, then the hydrogen temperature in the vehicle storage tank may rise to temperatures that may place in jeopardy the mechanical integrity of the storage tank itself. To address this problem, hydrogen fueling station operators follow one of two approaches: they either slow down the hydrogen fill-up rate, or they pre-cool the hydrogen below $0^{o}C$ so they can employ a high hydrogen fill-up rate. To better understand the tradeoffs in this decision making process we are proposing in this work a novel mathematical model, based on thermodynamic and transport phenomena fundamentals, which predicts the pressure and temperature of hydrogen in the vehicle storage tank during the fill-up process.

Conceptual Framework and Solution Approach

The system considered in this work consists of the vehicle storage tank and the valve that allows hydrogen to flow into the tank; figure 1 shows a system schematic.

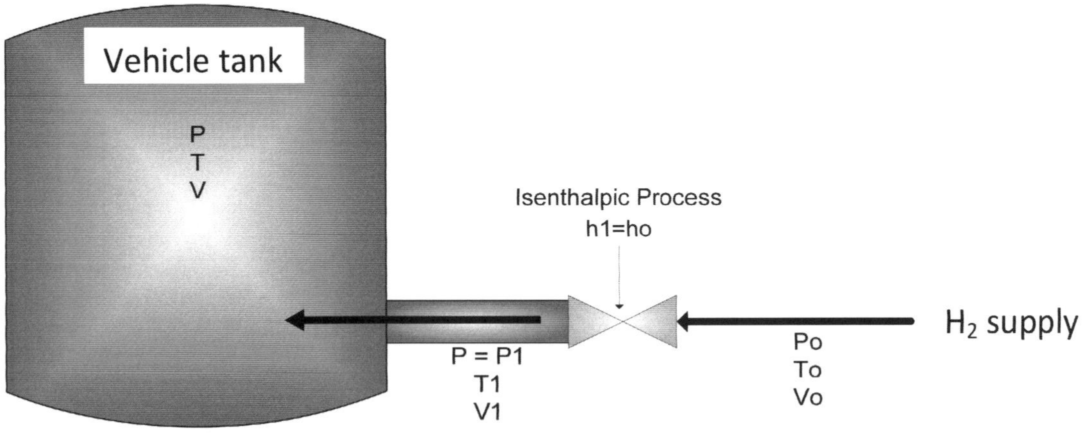

Fig.1. Illustration of the hydrogen fill-up process.

The system depicted in Figure 1 shows PVT conditions for three sections (from right to left): hydrogen supply from the fueling station, outlet of throttling valve/inlet of vehicle tank, and interior of vehicle tank. The hydrogen supply conditions (Po To Vo) and the fill-up mass flowrate are all assumed to be constant throughout the time interval of the fill-up. Then, the throttling valve outlet/ vehicle tank inlet conditions (P_1 T_1 V_1) are related to (Po To Vo) through an isenthalpic process that represents the valve operation. Finally, the vehicle tank conditions (P T V) are related to (P_1 T_1 V_1) through dynamic mass and energy balances on the hydrogen contained in the vehicle tank. These balances are developed under the assumption that the hydrogen pressure and temperature in the vehicle tank are uniform in space. In addition, it is assumed that the tank inlet and tank interior pressures (P_1 and P respectively) are equal, although the corresponding temperatures are not.

Our modeling effort for the hydrogen fill-up process is based on the premise that the hydrogen temperature rise phenomenon has two contributing mechanisms: heating by compression and heating by decompression (Joule-Thomson). The first mechanism suggests that the hydrogen temperature in the tank interior increases, as the pressure and hydrogen mass in the tank increase while the tank volume remains constant. The second mechanism suggests that as the hydrogen decompresses from the station storage pressure to the vehicle pressure by going through the isenthalpic valve, its temperature is increased. Indeed, the temperature of a real gas will increase or decrease as the gas passes through a valve and expands at constant enthalpy. The rise or drop in temperature of the gas is determined by the Joule-Thomson coefficient, shown in equation 1:

$$\mu = \left(\frac{\partial T}{\partial p}\right)_h \qquad (1)$$

Since the gas is undergoing a pressure drop by passing through the valve, its temperature will rise if the Joule-Thomson coefficient is negative, and will fall if the coefficient is positive. The "inversion temperature" of a gas is defined as the highest temperature at which the Joule Thomson coefficient is positive. For hydrogen this temperature is 205 K (-68 °C) and therefore at the fill-up conditions (temperature above 0 °C) the Joule-Thomson coefficient is always negative and the hydrogen heats up as it moves through the isenthalpic valve.

Quantifying the two aforementioned heating mechanisms requires that a thermodynamic model be employed for the hydrogen gas. At the considered pressures (350 bar to 700 bar depending on the station and vehicle generation) hydrogen does not behave as an ideal gas. As a result, we employ the Redlich-Kwong (RK) equation of state for non-ideal gases to describe hydrogen's thermodynamic behavior. The RK state equation allows the calculation, in a self-consistent manner, of hydrogen's internal energy and enthalpy given its temperature and pressure. These thermodynamic relations are then employed in the vehicle tank energy balance and in the energy balance across the valve. In the vehicle tank energy balance, a term is incorporated that accounts for the heat transfer between the tank and the environment. For this purpose, an overall heat transfer coefficient is estimated, accounting for free convection in the tank interior and surrounding air, and conduction through the tank wall made of carbon composite.

The resulting mathematical model consists of two ordinary, 1st order differential equations (mass and energy balances) and two algebraic equations (RK state equation and isenthalpic valve condition). This Differential-Algebraic-Equation (DAE) system is solved through the use of a 4th order Runge-Kutta approximation for the two ODE's in combination with a Newton-Raphson iteration for the computation of the tank inlet temperature (T_1) considering as known the tank pressure (P_1) from the previous time step. This hybrid Runge-Kutta, Newton-Raphson algorithm was implemented in MATLAB™.

Results and Discussion

The pressure and temperature in the tank interior and the valve outlet are shown in Figures 2 to 4. In Figure 2, the hydrogen pressure in the vehicle tank interior is shown as a function of time for several overall heat transfer coefficients. It is shown to be a monotonically increasing function of time, and a monotonically decreasing function of the overall heat transfer coefficient. The pressure curve obtained for the estimated heat transfer coefficient factor (UA=6.34 W/K) underestimates (not shown) typical data corresponding to the same average fill-up flowrate. In real life, the fill-up mass flowrate varies with time (it is high initially and low towards the end of the fill-up). Thus it should be expected that for low times the real data overestimate the model predictions. Towards the end, the slope of the model predictions is higher than that of the data because the real flowrate is now lower than the average flowrate used by the model. However, this slope change is not enough for the model to overestimate the real data at the end, given that initially it had significantly underestimated them.

Similar behavior is exhibited by the tank temperature shown in Figure 3. Additionally, the maximum temperature predicted for each heat transfer coefficient decreases as the heat transfer increases; the greatest temperature calculated was for zero heat transfer while the lowest corresponds to the highest heat transfer coefficient. Furthermore, the temperature switches from a linear shape to an exponential one with the enhancement of heat transfer.

The valve outlet hydrogen temperature is shown in Figure 4. As mentioned before, the hydrogen passes through an isenthalpic valve. During the first few seconds of the fill-up the tank inlet temperature rises to a maximum of 306K (33°C) due to the Joule-Thomson expansion experienced by the hydrogen. Then, as time progresses, the pressure in the vehicle tank interior increases thus reducing the pressure drop between the fed hydrogen and the vehicle tank; therefore, the valve outlet hydrogen temperature decreases as a function of time.

Additionally, the temperature rise in the vehicle tank interior is due to the hot hydrogen that underwent the decompression heat by going through the isenthalpic valve during the first minute of the fill-up. After this point, the temperature in the tank is greater than that of the valve outlet, and the temperature continues to rise due to the compression process. Therefore, the vehicle tank hydrogen temperature rise can be divided into two regimes depending on the dominant mechanism: a Joule-Thomson heating effect during the beginning of the fill-up and a heating by compression regime that takes over towards the end of the fill-up.

Conclusions

A novel mathematical model is proposed, based on thermodynamics and transport phenomena fundamentals, that aims to capture the hydrogen pressure and temperature during a hydrogen vehicle fill-up process. The model predictions are shown to slightly underestimate real fill-up data. A plausible explanation for this underestimation is that the model employs a constant fill-up mass flowrate and hydrogen supply pressure, while real life fill-ups occur under variable flowrate and pressure conditions. Our future research efforts will focus on removing these process model limitations.

Acknowledgment

This work was funded by the National Science Foundation NSF-CBET 0829211 and NSF-CBET 0943264 grants to which we are gratefully thankful. Also, we want to thank contributions of the UCLA Center for Excellence in Engineering and Diversity (CEED) for the realization of this work.

Model Simulation Plots I

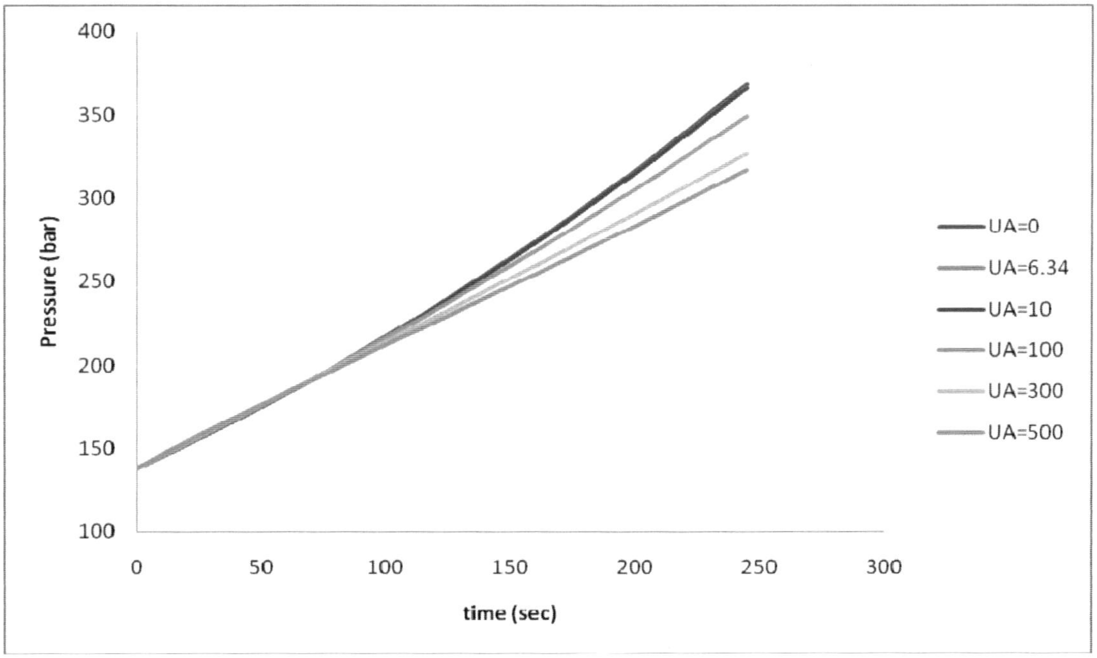

Fig. 2. Vehicle tank interior hydrogen pressure as a function of time for different overall heat transfer coefficients.

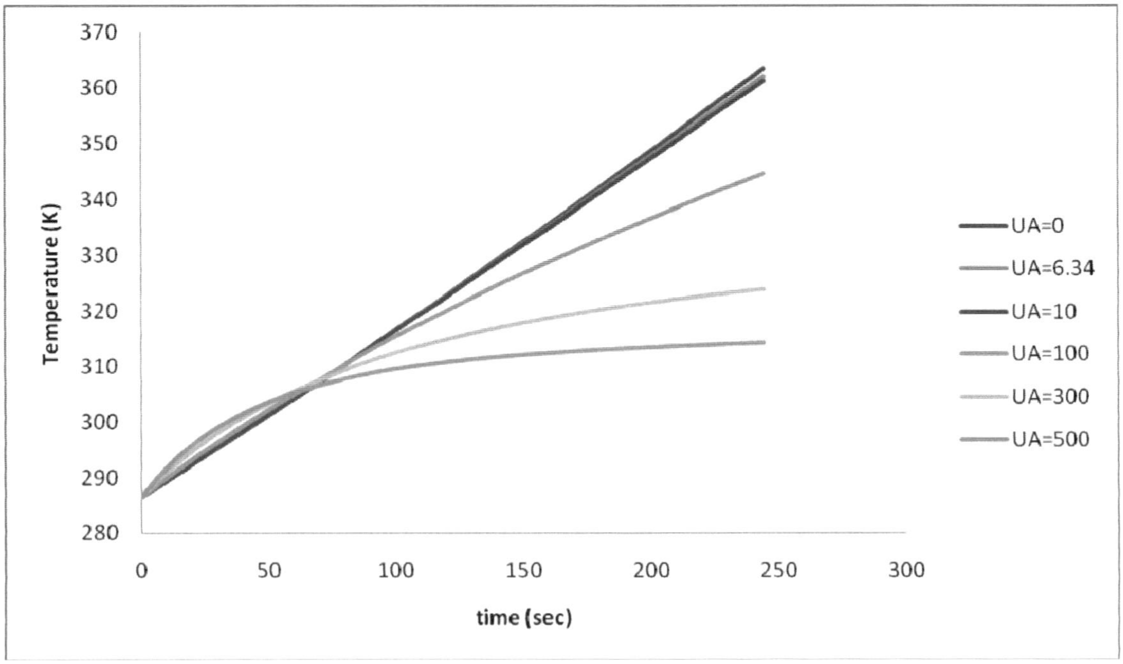

Fig. 3. Vehicle tank interior hydrogen temperature profile as a function of time for different overall heat transfer coefficients.

Model Simulation Plots II

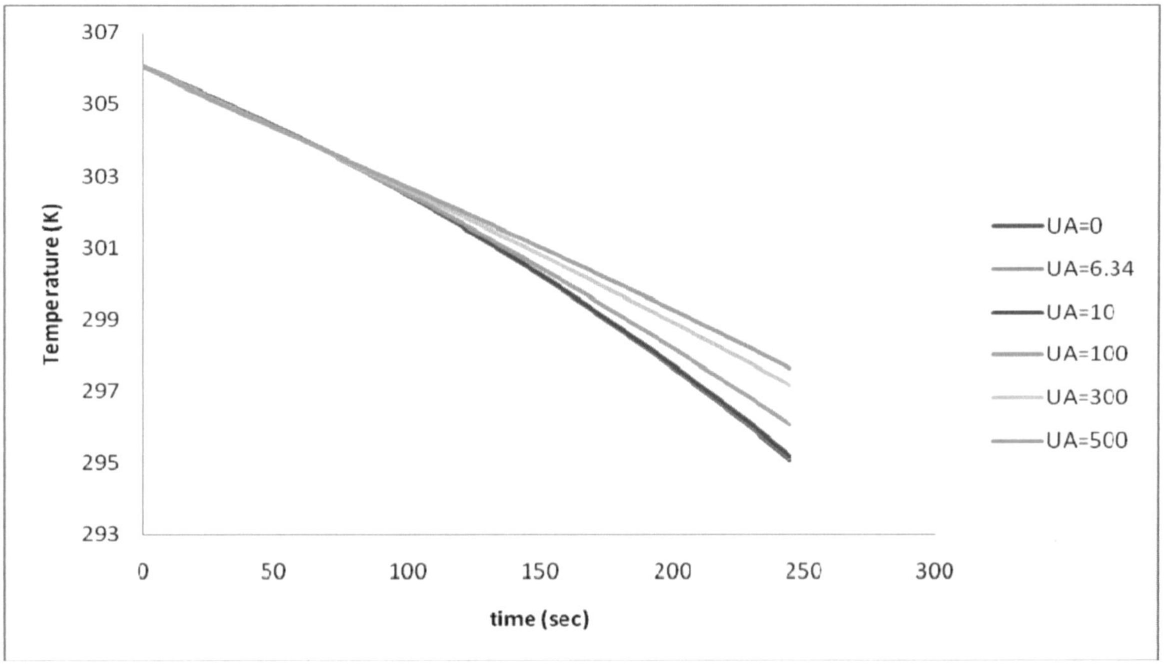

Fig. 4. Valve outlet hydrogen temperature as a function of time for different overall heat transfer coefficients.

(117a) Post-Combustion Carbon Capture Research Progress through the DOE/NETL Existing Plants Program

Timothy E. Fout, Existing Plants Division, National Energy Technology Laboratory, Morgantown, WV

Jared Ciferno, U.S. Department of Energy, National Energy Technology Laboratory, Pittsburgh, PA

Lynn Brickett, Existing Plants Division, National Energy Technology Laboratory, Pittsburgh, PA

Coal-fired power plants have made significant progress in reducing emissions of sulfur dioxide, nitrogen oxide, particulate matter, and now mercury (Hg), since the passage of the Clean Air Act. However, significant Congressional activity has been underway on legislation that would require a reduction in carbon dioxide (CO_2) emissions. A potential solution for the reduction of emissions from coal fired power plants is the capture and subsequent storage of CO_2 in a geological formation (CCS). DOE/ NETL's Existing Plants Program is focused on developing advanced CO_2 capture and compression technologies to aid in the development of the CCS process. The focus of the Existing Plants Program is driven by the fact that existing pulverized coal (PC) power plants produce a sizeable portion of current CO_2 emissions from all fossil-fuel-based sources, and that only about six gigawatts of the existing coal-fired electricity generating fleet is projected to retire by 2030. Current analysis indicates that state-of-the-art CO_2 capture technologies are not cost-effective technologies for capturing CO_2 from PC-based power plants. Also, these technologies have not been utilized on the scale required for CCS from a coal-fired power plant. The technologies being researched through the Existing Plants program aim at achieving a 90% capture rate while achieving no more than a 35% increase in the cost of electricity. This paper will give an update on the status of the program and recent progress of its post-combustion research projects.

See more of this Session: Post-Combustion Techniques for Capture/Separation of Carbon Dioxide I
See more of this Group/Topical: Topical 2: Advanced Fossil Energy Utilization

(117b) Vapor-Liquid Equilibrium (VLE) Study of CO2 with MEA and Ammonia Solvents

Donald Johnson, Power Generation and Utility Fuels, The Center for Applied Energy Research / University of Kentucky, Lexington, KY

Zhe Lu, Power Generation and Utility Fuels, The Center for Applied Energy Research / University of Kentucky, Lexington, KY

Joseph E. Remias, Power Generation and Utility Fuels, The Center for Applied Energy Research / University of Kentucky, Lexington, KY

Kunlei Liu, Power Generation and Utility Fuels, The Center for Applied Energy Research / University of Kentucky, Lexington, KY

James K. Neathery, Power Generation and Utility Fuels, The Center for Applied Energy Research / University of Kentucky, Lexington, KY

In the carbon capture and sequestration (CCS)scenario, vapor liquid equilibrium (VLE) curves represent the theoretical relationship between local gaseous CO2 vapor pressure and the respective carbon loading in solution for a given operating temperature. It determines the maximum carbon content in the lean solution to achieve 90% removal efficiency in the absorption process, and the minimum carrier gas flowrate required in the stripper for solvent regeneration at operating temperature. Unfortunately, the equilibrium between liquid and vapor phases can be very difficult to accurately quantify with simulation calculations, which use a combination of physical and thermodynamic equations along with post-experimental data for reference. Simulations can often yield results which are not a true reflection of what is observed in reality. In order to measure the VLE accurately, a CSTR equilibrium cell in conjunction with an on-line gas-sampling gas chromatograph (GC) for analyzing the vapor phase composition at equilibrium have been assembled. This apparatus will allow the engineer to test various solvents and determine the optimum operating conditions suitable for CCS. In this research effort, we report on initial gas equilibrium results that will be useful for designing CO2 scrubbers and associated stripping units using various aqueous ammonia and amine-based solvents. The results are reported at different CO2 loadings, temperatures, and pressures to obtain "true" vapor-liquid equilibrium curves for use in analysis and process design.

See more of this Session: Post-Combustion Techniques for Capture/Separation of Carbon Dioxide I
See more of this Group/Topical: Topical 2: Advanced Fossil Energy Utilization

Stripper Configurations for CO_2 Capture By Aqueous Amines

David H. Van Wagener, The University of Texas at Austin, Austin, TX; Dr. Gary T. Rochelle, The University of Texas at Austin, Austin, TX

Introduction

Amine-based absorption is considered an excellent option for CO_2 capture from coal-fired power plants. Compared to other options like oxycombustion and separation with membranes, absorption with amine solvents is advantageous because it has a relatively lower energy requirement for carbon capture and can be installed on the tail end of pre-existing coal plants. The majority of energy requirement for this process stems from solvent regeneration in the stripping section. A simple stripper is commonly used, but more complex configurations can help reduce the required energy input. It has been hypothesized that there is a relation between configuration complexity and performance that can be correlated. Complex configurations are desirable to improve stripper performance because of their ability to reduce driving forces. Fundamental work by Leites (2003) showed the benefit of reducing driving forces for generic chemical processes. Driving forces surface in several forms: temperature driving forces, mass transfer driving forces, and chemical reaction driving forces. Typical strippers for this application have very large driving forces in various locations. For example, the section of packing toward the bottom near the reboiler usually has a very large driving force, and the rest of the packing is nearly pinched, accomplishing the last of the reaction and mass transfer to approach equilibrium. Introducing more complexity to the by means of splits, recycles, and multiple pressure stages, the existing driving forces can be reduced to cut down on total exergy loss. In addition to reducing driving forces, more complex configurations with multiple pressure stages can improve the overall performance of the stripper and CO_2 compressor by collecting high-pressure CO_2, thereby reducing the total compressor work. High pressure stripping is beneficial to improve the selectivity of CO_2 while collecting at high pressure. Ambient pressure stripping is helpful to achieve a desired lean loading at lower temperatures, which can avoid excessive thermal degradation of the solvent.

Methods and Results

Complexity Modeling

Since complexity is a broad undertaking, the analysis started at the most simple separation method, a single heated flash, and progresses systematically to more complex configurations (Rochelle 2009). The mathematical definition of complexity is still loose, but it is expected that the number of process units in the configuration is roughly related to a complexity "value". The complexity is calculated as the total number of major pieces of equipment, excluding typical pumping and compression equipment. For example, a single stage heated flash has a value of 2 (a heater and separation vessel), and a simple stripper has a value of 3 (a heater, vessel, and packing). The complexity value of a configuration is decreased if an intercooler is eliminated by integrating a compression stage in the flowsheet. This is the case in the 2-Stage Multi-Pressure configuration. It is expected that more complex configurations will improve performance but have a diminishing return at higher complexity. It is also expected that different configurations with similar complexity values will have different performance according to how efficiently the process units are arranged. A systematic justification for the arrangement of process units as it relates to performance may be possible.

The analysis so far has focused on configurations using one, two, and three pressure stages. Each configuration was specified to run isothermally, except for special cases which included adiabatic sections. 9 m MEA was chosen as the solvent for this study due to the proven robustness of MEA. Additionally, 9 m MEA has improved solvent capacity over the industry standard, 7 m MEA. A thermodynamic model was developed for MEA by Hilliard (2008) which represents 9 m MEA fairly well; this Aspen Plus model using the electrolyte-NRTL method for MEA was used for this work. The different configurations can be separated into subcategories: multi-stage flash, and double matrix, and columns. The analysis began with the most simple separation method, a 1-stage flash, and built upward in complexity and total number of pressure stages. Each stage had a number of options: flash vs. packing, heated vs. adiabatic, and recompression of the vapor to the prior stage. The configurations that have been modeled are described in Figure 1 – 5. An additional configuration not shown is a stripper with an adiabatic lean flash. However, this flowsheet stems directly from the 2-stage multi-pressure configuration; the adiabatic lean flash configuration is a multi-pressure flowsheet with packing and the reboiler in the top section and an unheated flash in the bottom section.

Several of these configurations were simulated using multiple run specifications. The 2-stage flash was simulated for three cases: both stages heated isothermally, only the first stage heated, and only the second stage heated. These options were explored to determine the effectiveness of isothermal stripping. Next, the double matrix configuration was tested with a variety of options. First, it was simulated using only 3 flash sections as shown in Figure 5. This first case was also run with a uniform 20% split of the rich feed to the low-pressure flash. This value of split was used because it was found to be an optimum value in previous work. In the next case, this double matrix flash configuration was optimized for rich split at each lean loading. Finally, the double matrix configuration was also simulated with a section of packing in place of the adiabatic flash on the top of the low-pressure section. Another configuration not shown is the 3-stage flash, which follows the sequential progression of the multi-stage flash, similar to the 1- and 2-stage flash configurations. The vapor streams fed to the multistage compressor were pre-cooled to 40°C to knock out water vapor. Additionally, there was intercooling to 40°C between each stage of the compression trains.

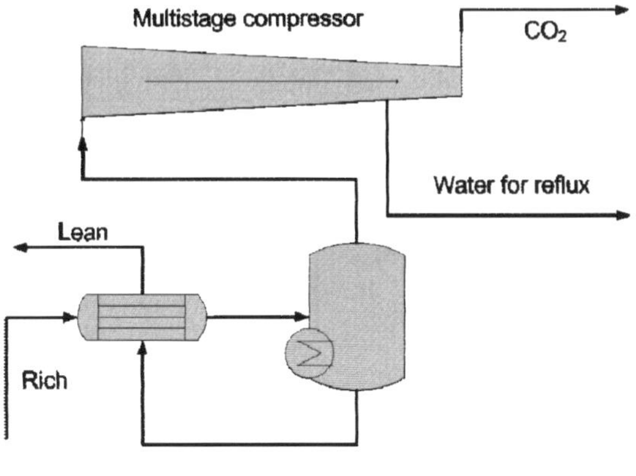

Figure 1. 1-Stage Flash

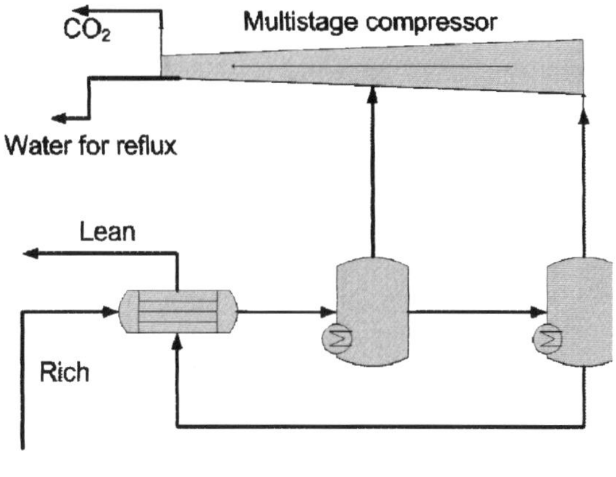

Figure 2. 2-Stage Flash

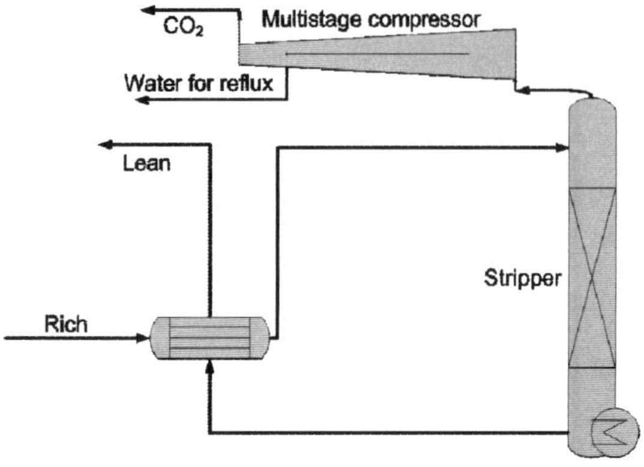

Figure 3. Simple Stripper

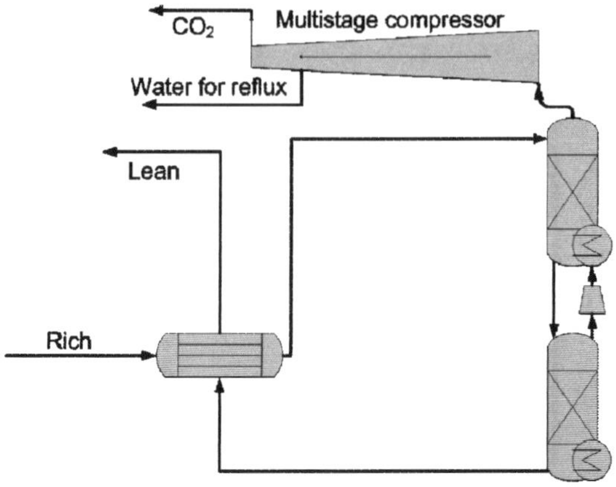

Figure 4. 2-Stage Multi-Pressure

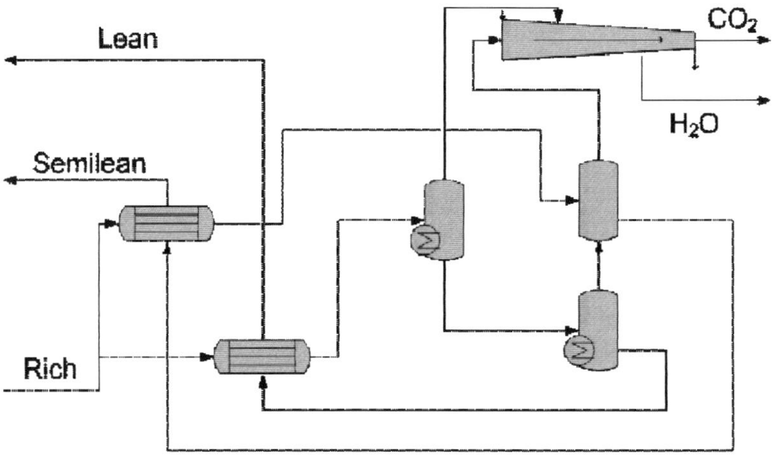

Figure 5. Double Matrix with Equilibrium Flashes

The simulations were performed with as many variables held constant as possible. With this goal in mind, the stripped CO_2 was compressed to 15 MPa using 7 stages in all simulations. This compression also included precooling and intercooling to 40°C with water knockout in each cooler. The rich pumps achieved the vessel pressure for a given configuration, and it also accounted for an additional 50 kPa for possible frictional losses as well as 250 kPa for gravitational losses if a packed column was used in exchange for a flash vessel. A constant rich loading of 0.5 was used, which corresponds approximately to a CO_2 partial pressure of 5 kPa in the solvent at 40°C, a typical rich end temperature for the absorber. A 5°C cold side approach was specified on all heat exchangers. A 10°C approach was specified on the steam reboilers. Equivalent work was calculated as follows:

$$W_{eq} = \sum_{i=1}^{n_{reboilers}} 0.75 * Q_i \left(\frac{T_i + 10K - T_{sink}}{T_i + 10K}\right) + W_{pumps} + W_{comps}$$

where Q_i and T_i are the duty and temperature for reboiler i. The equivalent work represents the total electrical penalty on a power plant assuming that electricity could be extracted from steam at a 75% Carnot efficiency rate. The lean loading was optimized to minimize total equivalent work for each configuration. The equivalent work was evaluated for lean loadings ranging from 0.2 to 0.45. The optimum lean loadings were all above 0.35, but the performance was evaluated down to lean loadings of 0.2 because many industrial users prefer these low lean loadings to reduce absorber capital cost.

Since isothermal operation should produce the most efficient thermodynamic results, all reboilers and heaters in each run were specified to have equal temperatures. The analyses were run for 100°C, 110°C, 120°C, and 130°C. Unless stated otherwise, the results presented in this report will be only for the 120°C runs. This temperature is expected to be the highest feasible operating temperature for MEA due to elevated thermal degradation rates above 120°C. The results from the different categories of configurations are charted below in Figure 6 - Figure 10. Additionally, the minimum equivalent work for each configuration is documented in Table 1. This table also lists the lean loading at the minimum equivalent work as well as the complexity rating, calculated by the basic method presented earlier.

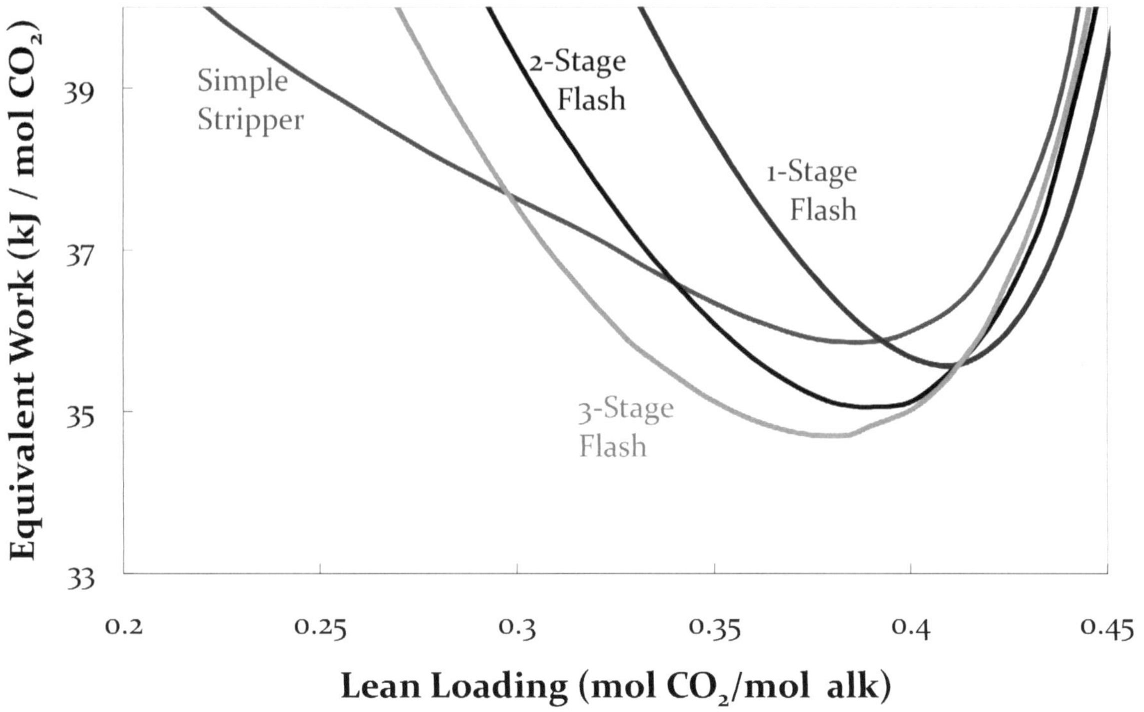

Figure 6. Multi-Stage Flash Results. 9 m MEA, 0.5 rich loading, 120°C reboilers, CO_2 compression to 150 atm, evenly distributed vapor generation per stage on mole basis.

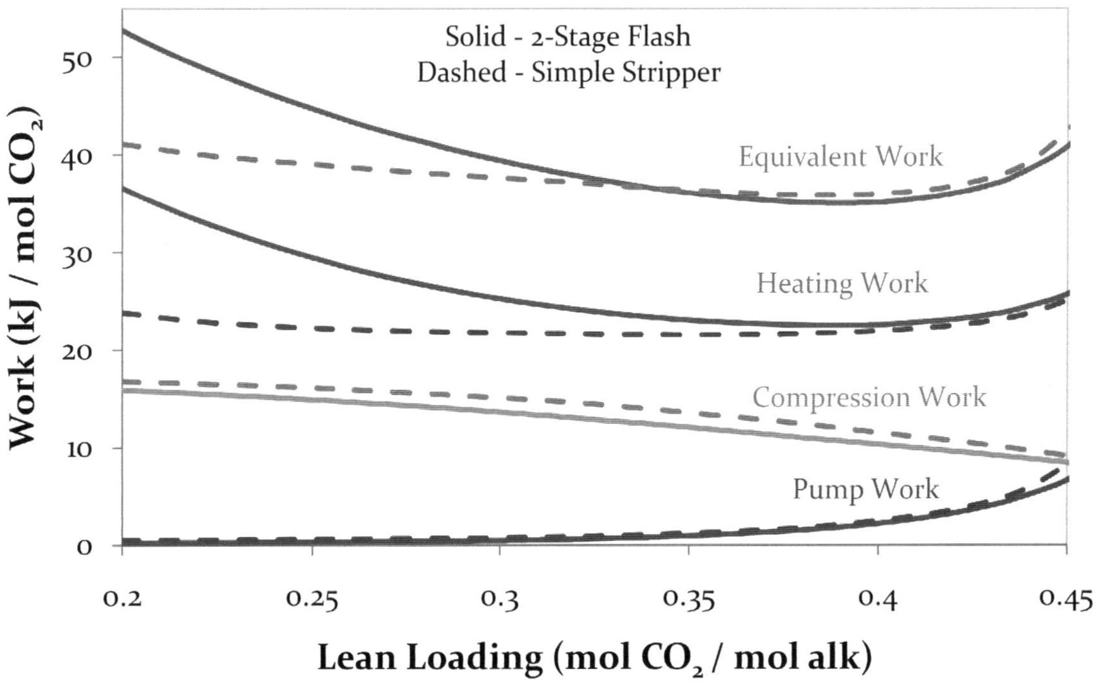

Figure 7. Contributions to Equivalent Work for 2-Stage Flash and Simple Stripper

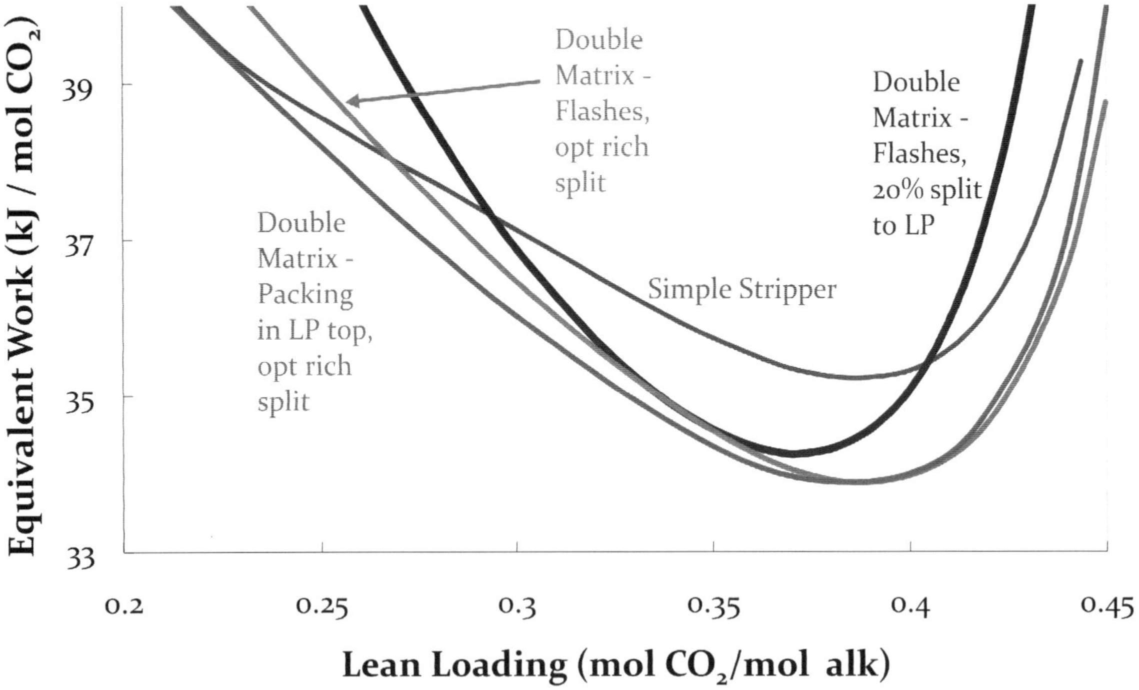

Figure 8. Double Matrix Results. 9 m MEA, 0.5 rich loading, 120°C reboilers, CO_2 compression to 150 atm, evenly distributed vapor generation per stage on mole basis.

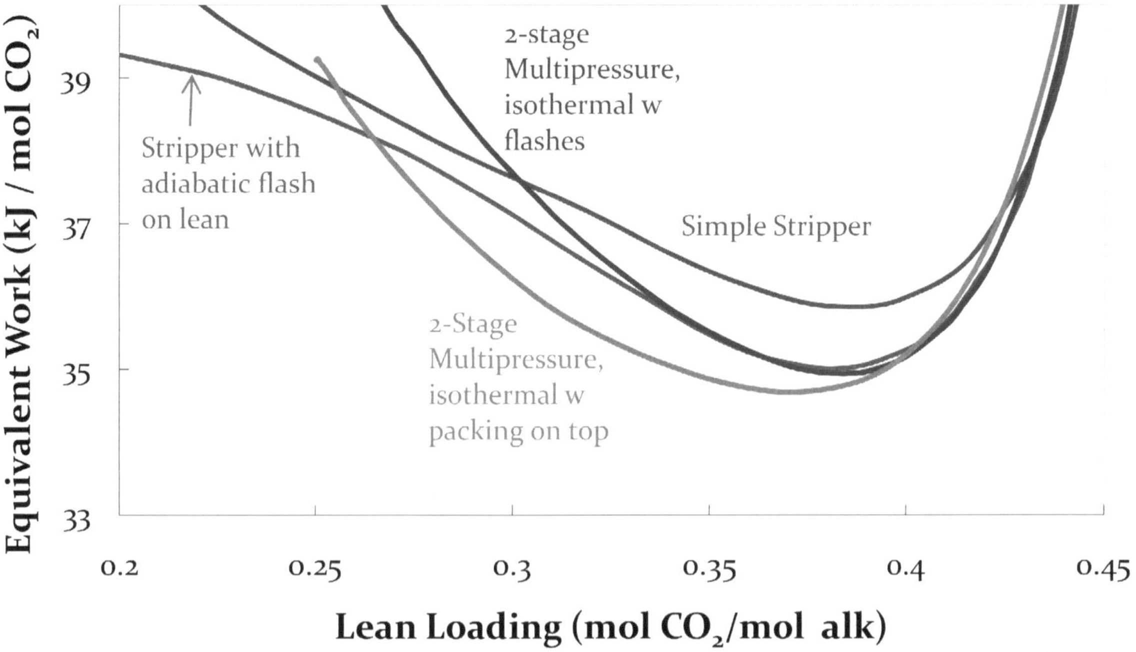

Figure 9. Column Results. 9 m MEA, 0.5 rich loading, 120°C reboilers, CO_2 compression to 150 atm, evenly distributed vapor generation per stage, mole basis (except Multipressure - 1.8 compression ratio).

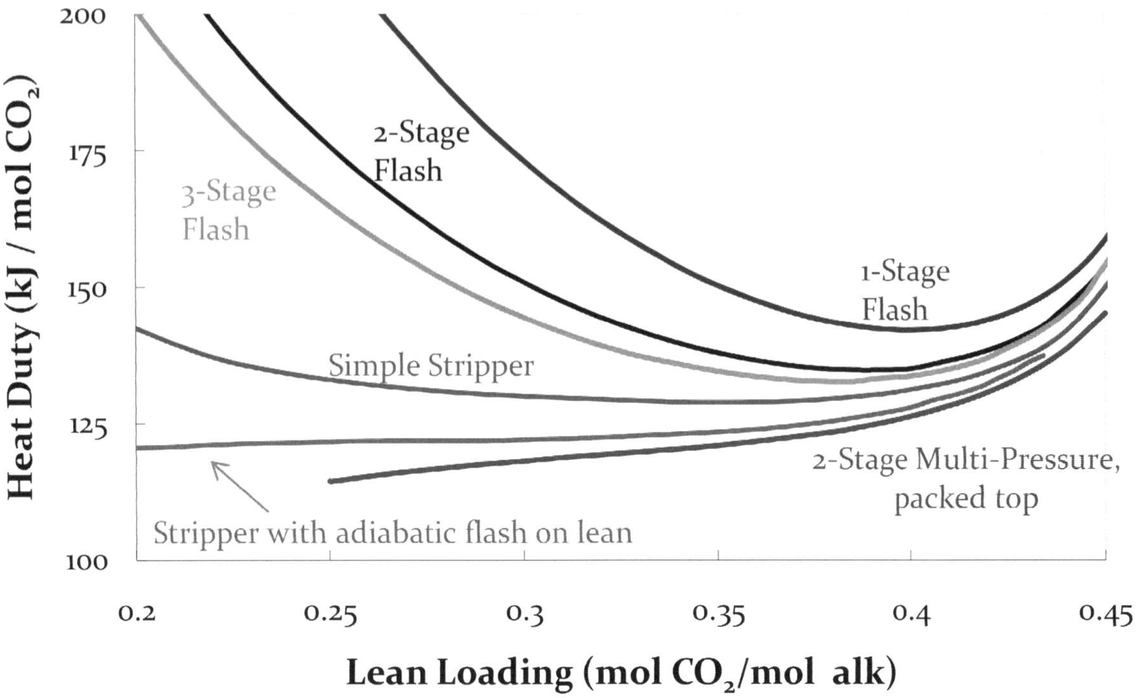

Figure 10. Heat Duties for Selected Configurations. 9 m MEA, 0.5 rich loading, 120°C reboilers, CO_2 compression to 150 atm, evenly distributed vapor generation per stage, mole basis, except multi-pressure.

Table 1. Results Summary for All Configurations

Configuration	Process units	Equivalent Work kJ/mol CO_2	Optimum Lean Loading
1-Stage flash	2	35.6	0.410
Simple Stripper	3	35.2	0.383
Stripper with adiabatic lean flash	3	35.0	0.387
2-Stage flash, 1st adiabatic	3	35.4	0.412
2-Stage flash, 2nd adiabatic	3	35.2	0.400
2-stage Multi-pressure, flashes	3	34.9	0.385
2-Stage flash	4	35.1	0.390
2-stage Multi-pressure, packed top	4	34.7	0.370
Double matrix, flashes	5	33.9	0.390
3-stage flash	6	34.7	0.380
Double matrix, packed LP top	6	34.0	0.370

The plots above demonstrate a clear improvement in stripper performance as complexity increased. This point is especially apparent in Figure 6, showing the difference between the various multi-stage flash configurations. Not only did the energy requirement decrease with more flash stages, but the optimum lean loading also decreases. This lower lean loading would be beneficial for the absorber to improve its performance and decrease the absorber packing height. The more complex configurations shown in the other figures decrease the energy requirement even more. In Table 1 the general trend of decreasing equivalent work with increasing complexity is displayed, though the trend did not fit a direct correlation with the number of process units. The improvement in performance also depended on two factors: use of vapor recompression and replacement of flashes with packing. Recompression of the low-pressure vapor improved performance by condensing some of its stripping steam to vaporize CO_2 at the higher pressure. Unlike vapor recompression, inserting packing sections in place of equilibrium flashes did not improve performance as much as expected. The effect of this addition was seen directly in the double matrix and 2-stage multi-pressure configurations. In the case of the double matrix, the baseline configuration with three equilibrium flashes was altered by replacing the top flash of the low-pressure section with a packed column. Adding packing in this location was expected to have the greatest effect on performance because a large amount of stripping steam was produced in the bottom flash of the low-pressure section, and the stripping steam required an adequate number of transfer units to fully condense and vaporize CO_2. The other flashes included reboilers which mostly eliminate the need for transfer units. Similarly, in the 2-stage multi-pressure configuration, packing was inserted in the high-pressure stage to better reach equilibrium with the inlet rich to reduce the amount of stripping steam in the vapor before it was cooled and compressed. As seen in Figure 8 and Figure 9, including packing in these two configurations had no effect at high lean loadings where the CO_2 capacity was low. However, packing improved performance in both configurations at low lean loadings when an increased number of transfer units was beneficial to achieve the increased amount of CO_2 liberation per unit solvent.

This analysis also demonstrated the significance of evaluating performance by equivalent work as opposed to heat duty. An analysis using heat duty neglected the effect of temperature on the value of steam. This heat duty analysis also overlooked the pumping and compression contributions to overall energy requirement. An example of this difference could be seen when comparing the 2-stage flash with the simple stripper in Figure 7. The 2-stage flash has a lower equivalent work but higher heat duty than the simple stripper. The improvement of the 2-stage flash over the simple stripper stemmed from reduction in compression work due to collection of CO_2 at high pressure in the first flash, so the heat duty analysis does not recognize this benefit. Additionally, Figure 10 shows that the optimum lean loading when evaluating heat duty was seemingly flatter than when evaluating equivalent work, which can be explained by the heat duty analysis neglecting the increased value of steam at high temperatures encountered when stripping to low loading.

Conclusions

Various stripper configurations with different levels of complexity have been analyzed in Aspen Plus® using the MEA model developed by Hilliard. 9 m MEA was the solvent. The multistage flash configurations demonstrated the improved performance with more pressure stages. More pressure stages typically yielded better performance due to more reversible operation. Replacing flash stages with packing sections has repeatedly demonstrated better performance at low loadings, though the improvement in performance at optimum loadings was generally insignificant.

The baseline simple stripper required 35.2 kJ/mol CO_2. The double matrix configuration performed best, and its optimum performance was nearly identical regardless whether packing was used at the top of the low-pressure section. The 3-stage flash and 2-stage multi-pressure configurations also performed well, but the multi-pressure had a significantly lower complexity level. The double matrix required 34.0 kJ/mol CO_2 with a complexity rating of 6, the 3-stage flash required 34.7 kJ/mol CO_2 with a complexity of 6, and the 2-stage multi-pressure required 34.7 kJ/mol CO_2 with a complexity of 4.

References

Hilliard, Marcus D. *A Predictive Thermodynamic Model for an Aqueous Blend of Potassium Carbonate, Piperazine, and Monoethanolamine for Carbon Dioxide Capture from Flue Gas.* Ph.D. Dissertation, The University Of Texas at Austin, 2008.

Leites, I L, D A Sama, and N Lior. "The theory and practice of energy saving in the chemical industry: some methods for reducing thermodynamic irreversibility in chemical technology processes." *Elsevier*, 2003: 55-97.

Rochelle, GT at al. "CO2 Capture by Aqueous Absorption, Third Quarterly Progress Report 2009." *Luminant Carbon Management Program.* The University Of Texas at Austin, 2009.

(117d) Enhancing Post-Combustion CO2 Capture in Power Plant – A Catalytic Approach

Reynolds A. Frimpong, Center for Applied Energy Research, University of Kentucky, Lexington, KY

Quanzhen Huang, Center for Applied Energy Research, University of Kentucky, Lexington, KY

Joseph E. Remias, Center for Applied Energy Research, University of Kentucky, Lexington, KY

James K. Neathery, Center for Applied Energy Research, University of Kentucky, Lexington, KY

Kunlei Liu, Center for Applied Energy Research, University of Kentucky, Lexington, KY

Improved solvent reaction kinetics and carbon capacity continue to be a major research interest for the scrubbing of CO2 in post-combustion processes. The DOE has set a new target for post-combustion CO2 capture of less than a 35% increase in the cost of electricity, while capturing at least 90% of the CO2 released during the combustion of fossil fuels in existing coal-fired power plants. In order to meet the DOE target, it will be necessary to promote the CO2 absorption rate in aqueous capture solutions. This work will discuss the use of catalysts to promote the reaction rate. Some current research efforts are targeted at developing new solvents or enhancing the absorption capabilities of some of the solvents already in use such as amine-based, ammonium-based, and alkali-carbonates. The use of catalytic additives in these solvents is one approach to promoting CO2 capture.

In this work, we describe the synthesis, characterization, and use of low cost non-toxic additives such as zeolites for CO2 capture rate enhancement. Ion-exchange reactions and metal impregnation were carried out on these additives to improve their catalytic properties. Results have shown that some zeolites provide catalytic enhancement in CO2 scrubbing when added to aqueous ammonium-based solutions.

See more of this Session: Post-Combustion Techniques for Capture/Separation of Carbon Dioxide I
See more of this Group/Topical: Topical 2: Advanced Fossil Energy Utilization

(117e) Design of a Continuous Foam-Bed Reactor System for CO2 Removal From Power-Plant Exhausts

Amit A. Gaikwad, Department of Chemical Engineering, Indian Institute of Technology, Delhi, New Delhi, India

Ashok N. Bhaskarwar, Department of Chemical Engineering, Indian Institute of Technology, Delhi, New Delhi, India

With renewed emphasis on reducing the carbon footprint in view of the United Nations Climate Change Conference in Copenhagen, reducing the greenhouse gas emissions is of prime importance. The average concentration of CO_2, the prominent of greenhouse gases, in the atmosphere today is about 385 ppm, which was about 280 ppm before the industrialization started. Intergovermental Panel on Climate Change (IPCC) says that the Earth's average temperature has risen by 0.74 degrees in the period from 1906 to 2005, and that the average temperature will continue to rise.

Various technologies are being developed to tackle this problem of CO_2 emissions. One such promising method is use of foam-bed reactors. The reactor offers very large gas hold-ups and low pressure drops, which is ideal for treatment of gaseous pollutants. This work focuses on treating CO_2 by reacting it with aqueous barium-sulfide solution in a continuous foam-bed reactor. The CO_2 required for this carbonation reaction could be obtained from smoke stack furnaces or power-plant exhausts, thereby reducing air pollution. H_2S gas is produced in the reaction, but it reacts faster with amines as compared to CO_2 and thus it can be removed with relative ease.

Some data for CO_2 removal using foam-bed reactors are available (Gaikwad and Bhaskarwar, 2006, 2007; Gaikwad et al., 2007) but they pertain to semi-batch mode of operation of the reactor. When continuous operation of a reactor is necessary, the conversion per pass is likely to be small, especially at high throughput rates of the liquid being processed. The conversion in the exit stream may, however, have to be quite high for the process to be acceptable from the point of view of meeting the required production level or environmental regulations. Multiple reactors will then appear, among others, as a feasible option.

Subramanyam et al. (1999) have developed a design procedure for treatment of the effluent from Kraft paper mills using foam-bed reactor. The primary type in multi-reactor configurations is the reactors-in-series. For the present case of CO_2 removal using aqueous barium-sulfide solution (gas absorption with chemical reaction and desorption), the graphical approach used in the design of a foam-bed reactor system was similar to the one proposed by Subramanyam et al. (1999). The modified single-stage model of a foam-bed reactor (Gaikwad and Bhaskarwar, 2007) was applied to the system of N foam-bed reactors, all of equal size in series, operated continuously.

The design of the foam-bed reactor system was done for the following specifications:

i. Effluent flow rate = 100 tons/ day

ii. Concentration of aqueous barium-sulfide solution in the inlet stream = 0.56 k mol/m^3

iii. Desired conversion = at least 95%

iv. Gas-flow rate = 8 x 10^{-3} m^3/s

v. Concentration of carbon-dioxide gas in the inlet stream = 1 x 10^{-2} k mol/m^3

A design chart has been prepared in the form of figure 1. In this chart, the height of a foam-bed reactor (H) has been plotted against the number of reactors required to be connected in series in order to achieve 95% conversion of the inlet liquid stream under the specified conditions (i-v). It is seen from the chart that the number of reactors and the height of each reactor are both smaller for a higher radius of the reactor (r_c). Such design plots have been made for different reaction velocities starting from 4300 m^3/k mol.s to 11000 m^3/k mol.s.

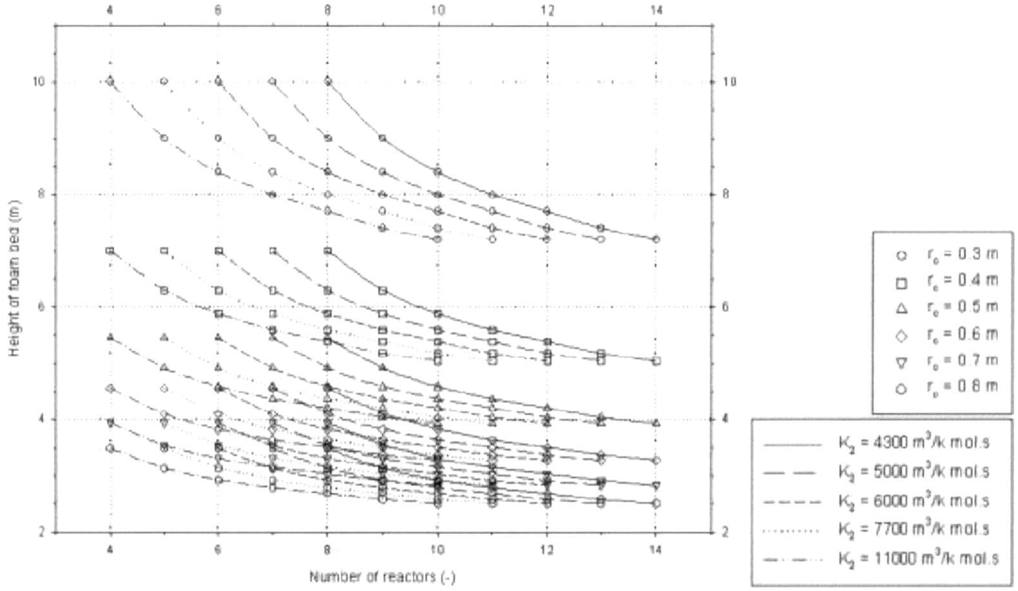

Figure 1. Variation of the height of foam bed with the number of reactors in (design chart I).

One can choose an aspect ratio (ratio of reactor height to reactor diameter) of say about 10, and find out the combination of H and r_c which gives a reasonably small number of reactors (say not greater than 10) for a given reaction velocity. Then, we have to turn to the second design chart shown in figure 2 wherein the average liquid hold-up in the reactor has been plotted against the number of reactors in series, again required for 95% conversion under the specified conditions (i-v). Having found out H and r_c in the previous step, one can now locate the point on the chart corresponding to the average liquid hold-up existing under the actual operating conditions by suitably interpolating for the corresponding value of the parameter r_c^2H, if not directly readable from curves shown. The number of foam-bed reactors actually required in series can then be directly read off from the corresponding abscissa. This number, if it is a fraction, will have to be rounded off to the nearest higher integer to ensure a conversion greater than 95%.

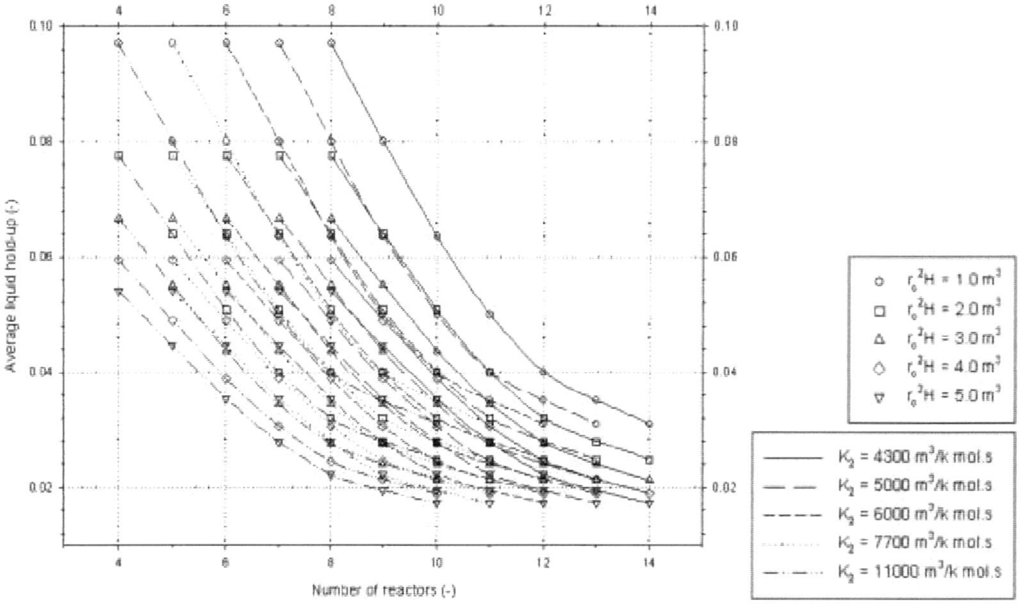

Figure 2. Variation of the average liquid hold-up with the number of reactors in series (design chart II).

Thus using these design charts one can obtain the dimensions and number of foam-bed reactors required for a continuous operation to remove CO_2. The design process covers broad range of reaction velocities giving a flexibility of choosing a reactant liquid other than aqueous barium-sulfide solution. Following the procedure outlined in this work, similar plots can be generated for different gas velocities, conversions, effluent flow-rates, and liquid and CO_2 concentration to obtain the required design parameters.

See more of this Session: Post-Combustion Techniques for Capture/Separation of Carbon Dioxide I
See more of this Group/Topical: Topical 2: Advanced Fossil Energy Utilization

(117f) An Incremental, Multi-Technology Approach to Retrofitting Existing Pulverized Coal Power Plants for Carbon Capture

David C. Miller, Collaboratory for Process & Dynamic Systems Research, National Energy Technology Laboratory, U.S. Department of Energy, Morgantown, WV

Global climate models indicate that acting rapidly to reduce carbon dioxide emissions will provide greater benefits to the environment than delaying action; however, new technologies, currently under development, may be significantly less costly than those currently ready to be deployed. This tradeoff between time to deployment and cost makes it difficult to determine the best course of action. This paper will explore scenarios under which partial post-combustion capture is achieved with aqueous amine absorption in the near term and additional capture capability is achieved in the future through newer technologies, such as advanced sorbents or membrane processes. Results from process simulations will be presented to demonstrate the potential for a multi-technology approach to achieve synergistic reductions in the cost of carbon capture.

See more of this Session: Post-Combustion Techniques for Capture/Separation of Carbon Dioxide I
See more of this Group/Topical: Topical 2: Advanced Fossil Energy Utilization

(138a) A Two-Tiered Approach to Managing Greenhouse Gas Emissions in US Petroleum Refineries

Abhishek Pednekar, Process Engineering, WorleyParsons Resources and Energy, Monrovia, CA

Dennis Aguirre, Process Engineering, WorleyParsons Resources and Energy, Monrovia, CA

With increasing federal and state regulations aimed to control greenhouse gas (GHG) emissions, the question of profitable refinery operations have come to the limelight. The purpose of this study was to demonstrate our capability of using a two-tiered approach to help manage the GHG emissions and economically evaluate related business decisions. In a nut shell the two-tiered approach involves conducting refinery wide GHG emissions analysis and programming the findings into an operations planning model in order to manage emissions and maximize profitability. The approach described here forth is generalizable to other social and environmental constraints on a project and not limited to carbon emissions.

The first tier involves utilizing linear programming (LP) models. Linear programming is a technique for optimization of a linear objective function (eg. profit), subject to a set of linear constraints (viz. product specifications: RON, benzene content etc). LP models are routinely used by refineries for supply chain management and refinery planning operations. LP models are also effectively used in evaluating grassroots refinery configurations and revamp projects, in order to come up with the best operating plan to guarantee a maximum operating margin. LP models provide parameters for benchmarking efficient unit operations, depending on various supply/demand scenarios. Moreover, as we will demonstrate in our approach, LP models can be used for managing refineries' carbon emissions. In order to do that, we would need to incorporate the 'cost of carbon' and the permissible amount that can be generated on a daily basis as a linear constraint in the refinery LP model. If the refinery generates a higher quantity of carbon emissions than the maximum constraint, a resulting penalty would negatively impact the daily profit margin. Alternatively, a built in incentive would favor a minimum quantity of carbon emissions. The carbon emissions are directly linked to the type of crude oil that is processed in the refinery to the amount of fuel gas purchased and it varies widely within refinery units as shown by Andrawis et al, 2007. The LP model can, thus, be used as a tool for effectively managing carbon emissions on a daily operational level in the refinery while achieving optimum profitability.

Although the LP model is effective on an operational and planning level, a higher level approach is required for making business decisions regarding the viability of projects in an economic climate driven by restrictive environmental regulations. This leads us to the second tier in our approach which is the proprietary WorleyParsons methodology known as EcoNomics™. EcoNomics™ is used to evaluate project profitability by taking into consideration all the socio-economic and environmental parameters that are ignored in a traditional economic evaluation. The presentation will show the methodology and examples of how EcoNomics™ can be used to make critical business decisions by accounting for environmental costs like the cost of carbon. This approach can also be used to do sensitivity analyses (by using the EcoNomics™ Delta tool) to see how the bottom line of a project can be affected by the fluctuating cost of carbon over the duration of the project life cycle. It can also be used to economically evaluate options like carbon capture and storage (CCS), installing cogeneration plants etc. While, the second tier is an overall holistic approach of evaluating profitability of projects, it goes hand in hand by feeding critical information to the LP model, which is pretty much the tool used by refinery planners in their daily operations.

This two-tiered approach can be applied to existing projects or new projects in existing refineries and as mentioned earlier, it is generalizable to other environmental constraints (viz. total economic value (TEV) of water). The two-tiered approach can also be used to evaluate the potential integration of biofuels in existing petroleum refineries (ranging from blending of biofuels in transportation fuels to an integrated biorefinery) in order to meet future requirements.

See more of this Session: Greenhouse Gas Reductions through Industrial Efficiency Improvements
See more of this Group/Topical: Environmental Division

(138b) Effects of Nitrogen Ions Concentration On South African Fresh Water High CO2 Tolerant Microalgae Growth

Edmore Kativu, School of chemical and metallurgical engineering, University of the Witwatersrand, Johannesburg, Johannesburg, South Africa

Tonderayi Matambo, School of chemical and metallurgical enginering, University of the Witwatersrand, Johannesburg, Johannesburg, South Africa

Diane Hildebrandt, School of Chemical and Metallurgical Engineering, University of the Witwatersrand, Johannesburg, Johannesburg, South Africa

David Glasser, School of Chemical and Metallurgical Engineering, University of the Witwatersrand, Johannesburg, Johannesburg, South Africa

Karl Rumbold, School of Cell and molecular Biology, University of the Witwatersrand, Johannesburg, Johannesburg, South Africa

South Africa and the world at large are in a dire need to abate unprecedented CO2 levels that are believed to be causing unstable environment, phenomena generally called climate change. Microalgae are photosynthetic aquatic microorganism. They are promising to be a sustainable means of mitigating green house gas emissions and their biomass has valuable uses ranging from biofuels, organic fertilizers, stock feeds and natraceuticals. They are several factors affecting microalgae growth which includes nitrogen, phosphate carbon, light, temperature and micronutrients. A thorough understanding of the microalgae growth limiting factors is important to give information to engineers to design efficient green house gas emission abatement strategies. The amount of nitrogen ions required to achieve optimal growth rate and maximum biomass yields need to be investigated and quantified. In this study an unknown green microalgae was isolated from Johannesburg Zoo Lake, South Africa. Batch experiments to determine the effect of nitrogen ions species and concentration were carried in Phytotron growth chambers with fixed light intensity and temperature at 6000 lux and 25°C respectively for 12 hours photoperiod. Microalgae growth was assessed by measuring the optical density at 680nm. The nitrogen ion species used were NH4Cl, KNO3 and NH4NO3.Various nitrogen ions concentrations were prepared using the Beijerick medium in 1000ml Erlenmeyer flasks. The results show that the microalgae have no preferential utilization of nitrogen ion species. However the ammonium nitrate appears to have promoted slightly higher specific growth rates and dry biomass yields followed by ammonium chloride and lastly potassium nitrate though the difference was statistically insignificant. When the microalgae was cultured with a medium of nitrogen ions concentrations of 0 g/l, 0.02g/l and 0.03 g/l the overall biomass after 12 days was 76.05 mg/l, 130.17 mg/l and 179.90 mg/l respectively, on the 12th day 0.15g/l of nitrogen ions concentration were added to these nitrogen depleted cultures and there was a sudden increase in both growth rate and biomass yields in all experiments. Generally, an increase in nitrogen ion concentration resulted in increased specific growth rates and maximum dry biomass yields and 0.30g/l of nitrogen ion concentration was the maximum point and further increase in concentration resulted in a slightly reduced growth rates. When no nitrogen ions were added the specific growth rate and overall biomass yields were 0.10 per day and 76.05 mg/l respectively and reached reached a maximum of 0.55 per day and 624 mg/l respectively at nitrogen ion concentration of 0.3 g/ l. The obtained results indicate the importance of nitrogen ions in developing microalgae culturing system since nitrogen constitute at least 10% of the microalgae dry biomass.

See more of this Session: Greenhouse Gas Reductions through Industrial Efficiency Improvements
See more of this Group/Topical: Environmental Division

Using and Measuring the Combined Heat and Power Advantage

Tommy John PE, Tommy John Engineering, Inc.

Ameican Institute of Chemical Engineers, Spring National Meeting 2009

Abstract

Combined Heat and Power (CHP), also known as cogeneration, refers to the integration of thermal energy with power generation. CHP is a powerful energy conservation measure that has been identified as an important greenhouse gas reduction measure with net economic benefits. It complements other energy conservation measures. CHP can be used any place that heat is needed so it is used with a variety of applications, fuels, and equipment. There are ancillary benefits of CHP to the host site and the public including air quality, reliability, reduced water consumption, and economic development.

There is sometimes confusing information about the benefits of CHP because there is no universal standard for reporting the efficiency of CHP systems. This results in both overstatement and understatement of the benefits of CHP compared to other power generation systems. Efficiency is an important metric for use in assessing public policy and as a screening tool for economic analysis. Fuel Charged to Power (FCP) is the fuel, net of credit for thermal output, required to produce a kilowatt-hour of electricity. This provides a metric that is used for compaison to the heat rate of other types of generation and insight into the development of CHP projects that maximize economic and environmental benefits.

Biomass generation is generally less efficient than fossil fuel generation due to size and combustion characteristics, which means that there is more benefit from CHP because there is more waste heat available for recovery. An example is presented demonstrating that CHP significantly improves the economics and environmental benefits for biomass to power.

Combined Heat and Power

Combined Heat and Power (CHP), also known as cogeneration, refers to the integration of thermal energy with power generation. CHP plants are a major energy conservation technique because they are 70% to 90% efficient compared to a modern natural gas combined cycle power plant that operates at about 50% efficiency. CHP is an important greenhouse gas reduction measure that has net economic benefits. Shortage and higher costs of energy resources and the need to mitigate environmental impact make CHP a viable resource even though the plants are usually smaller and more complex. CHP can be used any place that heat is needed. The systems can be very large to serve the thermal load at a petroleum refinery or petrochemical complex, or small to match the heat load of a building, hospital, university campus, or school.

CHP currently provides about 8% of the power generated in the US, but there is potential for much more. Increased implementation of effective CHP will benefit not only the host site, but all energy consumers and the public with dramatic energy efficiency and air quality benefits in addition to improving the reliability of energy supply, reducing load on the transmission and distribution system, reducing water consumption, spurring economic development, and meeting potential carbon reduction requirements.

CHP Systems

CHP can use any fuel type; natural gas, biofuels, byproducts of agriculture and manufacturing, waste heat from manufacturing processes, and methane gas produced from landfills and anaerobic digesters. A variety of proven, reliable technologies are available to match any application including internal combustion engines, combustion turbines, steam turbines, and heat recovery systems. The most common fuel is natural gas. The following figure shows a natural gas combined cycle CHP plant. This arrangement, which often includes multiple gas turbines, provides the operator with maximum flexibility to supply thermal energy and optimize supply to the electric grid. Many plants include only some of the elements shown; for example a gas turbine and heat recovery steam generator (HRSG) only, or a boiler and back pressure steam turbine.

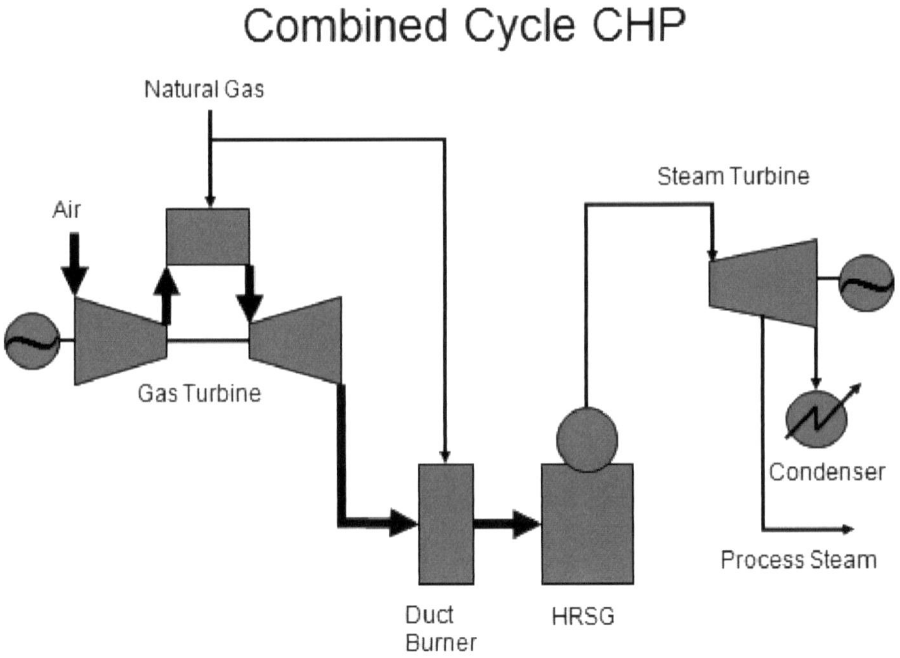

Energy Conservation

CHP complements other energy conservation measures to reduce global energy consumption and thus reduce combustion related emissions. An aggressive energy conservation program can yield savings of up to 30 to 40%. It starts with a thorough review of the process and then moves to optimizing equipment, controls, and utilities. Then and only then should CHP be considered. A CHP system configured to supply the optimum amount of energy can result in, on a global basis, additional energy savings equal or greater than those already achieved through conservation.

The benefits of CHP will only be realized if there is a clear understanding and measurement of those benefits for comparison to other energy options. Unfortunately there is sometimes confusing information about the benefits of CHP. There is no universal standard for reporting the efficiency of CHP systems, which can result in both overstatement and understatement of the benefits of CHP compared to other power generation systems. Efficiency is an important metric for use in assessing public policy and as a screening tool for economic analysis.

Fuel Charged to Power

The common metric for measuring the efficiency of thermal power plants is the Higher Heating Value (HHV) heat rate expressed as Btu/kwh (kj/kwh), which is the inverse of efficiency. A useful metric for CHP plants for comparison to other generation and for economic analysis is Fuel Charged to Power (FCP), the energy, net of credit for thermal output, required to produce a kilowatt-hour of electricity. For a plant using a gas turbine or engine including a combined cycle, a simple formula is:

$$FCP = \frac{Fuel - \frac{Heat}{Eff}}{Power}$$

Where Fuel is the total Btu/hr of fuel input including the prime mover and, if applicable, supplemental firing; Heat is the total thermal energy delivered to a process which is divided by the combustion efficiency of the alternative heat source; and Power is the kw of net power output. The net power output is the power produced by the combustion turbine or engine and a steam turbine (if applicable) minus parasitic load. The power output can be either electricity or mechanical power or both. Supplemental firing of combustion turbine exhaust will decrease FCP because duct burners use oxygen in the turbine exhaust and thus do not increase stack losses. In addition, the stack temperature actually decreases because the higher temperature differential inproves heat transfer rate, particularly in a single pressure level HRSG.

The FCP for a back pressure steam turbine is a special case. It is the conversion factor for kw-hr to Btu divided by the boiler combustion efficiency and the generator mechanical efficiency. Since all of the turbine exhaust is process heat, the incremental fuel is that needed to provide the power output.

$$FCP = 3412 / Blr\ Eff / Gen\ Eff$$

For example, if the boiler efficiency is 84% and the generator efficiency is 98%, the FCP is 4145 Btu/kwh. Note that the FCP is independent of the steam turbine isentropic efficiency. This does not mean that isentropic efficiency is irrelevant. Maximizing isentropic efficiency directly increases the amount of very efficient power that is produced saving energy and improving project economics. Maximizing inlet steam pressure and temperature and minimizing exhaust pressure also are key in increasing power output.

The output of cooling can be a source of confusion in determining CHP plant efficiency. While it is technically feasible to use Thermally Activated Cooling (TAC), such as absorption chillers, the inherently low Coefficient Of Performance (COP) compared to mechanical systems makes the systems inefficient. The energy displaced by the TAC is power instead of fuel, so the system is analogous to a combined cycle power plant where the prime mover output is enhanced by a bottoming cycle. In fact, for refrigeration loads over a few thousand tons, it is more efficient to use a condensing steam turbine driven mechanical chiller than an absorption chiller. The formula is:

FCP=Fuel/(Power+Cooling/COP)

For the units to be consistent, the cooling output has to be expressed in kw; the COP is for the alternative electric power driven chiller.

Natural Gas Generation

Natural gas fueled generation provides about 23% of the electric power in the US, but is often the incremental resource. In fact, in some regions natural gas is the marginal fuel for most of the hours of the year. The following table lists the heat rate (Btu/kwh) for natural gas generation alternatives and the typical FCP for CHP plants.

Steam Plants & Peakers	9,500-15,000
GE LMS 100 Simple Cycle	8,500
Combined Cycle (F-Class)	6,900
Combined Cycle (G-Class)	6,300
Gas Turbine CHP (FCP)	3,500-5,000
Gas Engine Jacket Water CHP (FCP)	3,900
Gas Engine/Turbine TAC (FCP)	9,600-14,500

Legacy steam plants and simple cycle gas turbines provide power at the time of peak demand. General Electric now markets the LMS100 gas turbine with a simple cycle heat rate of 8,500. There is a large fleet nationwide of F-class combined cycle plants, and the new G-class plants are even more efficient. CHP plants are able to use the improved gas turbine technology as will. The lowest FCP numbers represent the operation with significant supplemental firing, but a system with an efficient gas turbine and a well designed heat recovery system can operate with a FCP of 4,200 to 4,500 without supplemental firing. Smaller operations that can use heat at a temperature below 200 F can effectively use gas engines with jacket water and exhaust heat recovery. The FCP of a gas turbine or engine system providing cooling alone is competitive only with peak power.

In some cases CHP systems with TAC appears economic because it is blended with the delivery of thermal energy, because electric rates do not reflect the true marginal cost of generation, or because special incentives are provided based on thermal input instead of power actually avoided. It does little to provide real energy savings or environmental benefits compared to other natural gas generation options.

There are, however, applications where TAC makes sense. It has been demonstrated that hospitals can use CHP that include TAC effectively to remain in operation after a catastrophe like a hurricane destroying the electric grid. Since hospitals have a demand for thermal energy, a system with the flexibility to operate the TAC system only at peak times when the grid is available would be optimal. Another good application is a data center that requires reliable power and cooling.

The California Public Utilities Commission administers the Self Generation Incentive Program (SGIP) which provides generous subsidies for CHP and other measures that can be installed at customer sites. Performance problems within the program have been documented (1). There are multiple reasons, but certainly the lack of a realistic performance standard is a contributor. The efficiency standard for SGIP is the same as the one promulgated by the US Government for Qualifying (Cogeneration) Facilities (QF) under PURPA around 1980. The goal was to assure that a QF was at least as efficient as a combined cycle power plant, which then had a heat rate of about 8,000 Btu/kwh (compared to 6300 Btu/kwh with current technology). In addition, thermal input into a TAC system is counted instead of the equivalent power reduction. It is possible for a system that has a FCP of around 13,500 Btu/kwh to qualify for incentives.

Economics

Heat rate and FCP can be used to compare the economics of CHP with other power plants. As demonstrated in the following table, CHP is a very cost effective means of producing dispatchable base load power. The higher efficiency translates into costs for large scale CHP systems that are 30% lower than a natural gas combined cycle unit, which in itself is very competitive with other types of generation. Smaller CHP plants cost more because of the smaller scale and higher gas prices because they are typically connected to a local gas distribution company.

Fixed Charge Rate	15%
Line Loss	7%

	CC Gas	Large CHP	Small CHP
Unit Cost,$/kw	750	750	1700
Heat Rate, Btu/kwh	6750	4200	4200
Fuel,$/MMBtu	6.00	6.00	6.50
O&M,$/kwh	0.011	0.009	0.013
Capacity Factor,%	95%	95%	95%

TYPICAL POWER PRODUCTION COSTS,$/MWH

	CC Gas	Large CHP	Small CHP
Capital	14	14	31
Fuel	41	25	27
O&M	11	9	13
Generating Cost	65	48	71
Line Loss	5	0	0
Total	70	48	71

Biomass CHP

Biomass generation is generally less efficient than fossil fuel generation due to size and combustion characteristics, which means that there is more benefit from CHP because there is more waste heat available for recovery. The following example illustrates the value of CHP in reducing natural gas and water consumption as well as greatly improving project economics. The following table lists the basis for a biomass to energy plant using a steam cycle, the common technology for a plant of this scale. Natural gas is assumed to be the competing fuel for both power and heat.

•Biomass Fuel Consumption	100,000 tpy (dry)
•Capacity Factor	85%
•Biomass Heating Value (LHV)	7,866 Btu/lb
•Steam Conditions	850 psig, 800°F
•Steam Generation	116,000 pph
•Boiler Efficiency	73%
•Natural Gas	
–Cost	$8/MMBtu
–Combined cycle heat rate	7,000 Btu/kwh (49%)
–Boiler efficiency	83%
•Renewble Energy Credit (REC)	$10/MWH
•Power Credit	$81/MWH

If all of the steam is used in a condensing steam turbine, a net of over 10 MW of power is produced and over 100 million gallons per year of water is evaporated for cooling. If the steam is exhausted instead at 150 psig for process, the power output is reduced by about 60%, but 2.5 times as much natural gas is saved with no water use. The improvement in the project economics is dramatic as demonstrated in the following table. The capital investment and Operating and Maintenance expenses are assumed to be equal even though the CHP plant should actually be somewhat less expensive to build and operate since it does not have a cooling system. The project is assumed to qualify for a 30% renewable energy tax credit and the Biomass Crop Assistance Program.

	POWER ONLY	CHP
Net Power, KW	10,370	4,358
Exh Pressure, psia	1.0	165
Heat Rate/FCP, Btu/kwh	16,684	4,817
Water Use, mmgpy	107	-
Nat Gas Saved, mmcf/Yr	520	1283
Nat Gas Saved, $mm/Yr	4.3	10.7
Rec Value, $mm/Yr	0.9	0.3
Power Revenue, $mm/Yr	6.3	2.6
Thermal Revenue, $mm/Yr	0.0	8.9
EBITDA	0.7	5.5
Payout, Years	12.2	1.6

Conclusions

- CHP systems are most cost effective if designed to supply the optimum thermal energy use and at the lowest temperature.
- CHP systems should be designed to maximize heat recovery and efficient power output.
- Design for flexibility & reliability.
- Use the maximum unit size that will match the thermal requirements.
- Utilize any available by product or waste fuel.
- CHP enhances the economics of biomass to energy.

(1) Performance Assessment Of Cogeneration Systems In California; Kumar Chittory, George Simons, Brad Souza; Itron, Inc.; POWER2007-22114, Proceedings of ASME Power 2007

The Design of Ultra-Low NOx Critical Furnaces

Marco van Goethem [1], Iek Risseeuw and Simon Barendregt Technip Benelux, Zoetermeer, The Netherlands

Alessio Frassoldati, CMIC Department Politecnico di Milano, Milano, Italy

Prepared for: AIChE 2010, National Spring Meeting,
Group/Topical: Environmental Division,
session on Greenhouse Gas Reductions through Industrial Efficiency Improvements,
March 21-25, 2010, San Antonio, TX, USA.

Abstract
Environmental legislation such as the EU-IPPC directive, is becoming more and more stringent. As a furnace designer, Technip is faced with the challenge to meet ultra-low NOx emissions when designing or revamping ethylene cracking furnaces and steam reformers. This requires a detailed understanding of the fundamentals of fluid dynamics and combustion kinetics in such critical heaters. The paper deals with the fundamental modeling approach Technip has developed as a mandatory tool when designing critical furnaces.

Keywords: Ultra-low NOx, ethylene furnaces, CFD, NOx-Post Processor.

INTRODUCTION

Critical Furnaces are furnaces operating at elevated temperatures. Examples of such critical furnaces are ethylene furnaces and steam reformers. The process temperatures involved are typically above 850°C. Consequently, due to the high flue gas temperatures involved, those furnaces are sensitive to formation of Nitrogen Oxides. Nitrogen oxides formed during combustion are considered to be greenhouse gases as well a contributor to local pollution.
During recent years, the USA and the European Union and more recently also other countries, have imposed stringent NOx emission legislation on operators of industrial combustion installations. Other countries have recently imposed similar legislation.
The furnace and burner designer, when designing such critical furnaces, are facing a tough challenge to meet such imposed NOx emission ceilings. Burners are at the edge of what can be achieved in NOx reduction while maintaining proper burner performance such as flame stability and desired heat release pattern.
To reduce NOx emissions in ethylene furnaces, the designer has the choice to either reduce already formed NOx via flue gas treatment, or to minimize the formation of NOx at the source, which is the combustion process in the radiant section (the firebox) of the ethylene furnace. In either case, a proper understanding of the combustion kinetics and fluid-dynamics in the firebox chamber is required in order to arrive at a furnace design that combines stable and efficient combustion and that meets environmental NOx and CO emission constraints. This paper describes the fundamental kinetics mechanisms that are important in NOx

[1] Corresponding author: mvangoethem@technip.com

formation under combustion, as well as the combination of a rigorous combustion kinetics model with CFD (Computational Fluid-Dynamics) calculations.
The paper describes further the application of the combined CFD model to design a modern high-flux GK6^{TM} [2] ethylene furnace equipped with ultra-low NOx burners.

NOX AND COMBUSTION KINETICS

A kinetics model describing the oxidation of hydrocarbons up to kerosene and JP8 is used as basis. It contains some 250 species and 5000 reactions. Further, to describe the NOx formation and disappearance, the model contains 30 species and 260 reactions.
In combustion of methane and hydrogen containing fuels, NOx can be formed primarily by two reaction paths, the so-called Prompt NOx and the Thermal NOx mechanisms. (Figure 1). Other mechanisms involve the formation of NO via NNH and N_2O are of scarce importance in the applications discussed in this paper. As an example, N_2O mechanism becomes relevant at conditions such as those typically encountered in gas turbines, i.e. high pressure and very lean mixtures. The NOx mechanism and its validation are discussed in detail in [Frassoldati et al., 2003].

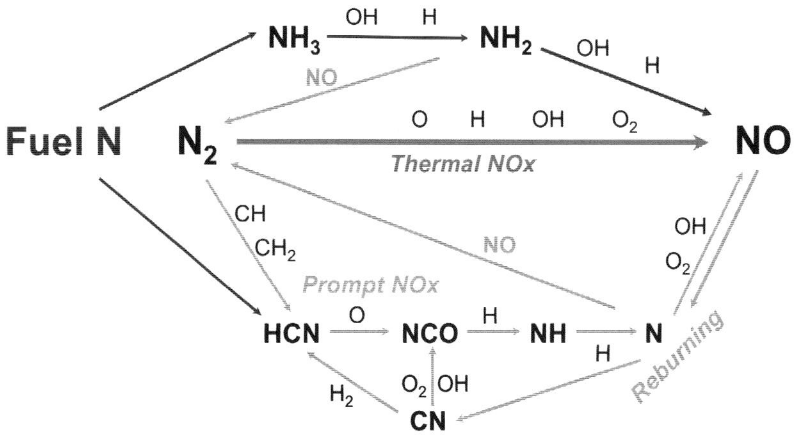

Figure 1: NOx formation mechanisms.

Thermal NOx

Thermal NOx is formed as a result of the oxidation of nitrogen to NOx through a reaction path that involves oxygen, hydrogen and hydroxyl radicals. Also the direct reaction between nitrogen and oxygen molecules contributes to the formation of this pollutant species via the thermal mechanism. High temperatures, high residence times and excess of oxygen favor thermal NOx. Avoiding local high flame temperatures and avoidance of too much excess air, can reduce thermal NOx formation.

Prompt NOx

Prompt NOx is the conversion of nitrogen through a radical reaction network, which involves hydrocarbon radicals formation with HCN as an intermediate. Prompt NOx is favored by excess of hydrocarbons, it is less temperature dependent than Thermal NOx and the reactions are fast compared to Thermal NOx.
Avoiding local excess of unburned hydrocarbons and keeping the flame lean of fuel can reduce Prompt NOx formation.

---

[2] GK6^{TM} is a trademark from Technip

Reduction (reburning) mechanism

Another mechanism is also important in NOx control, which is the reduction of already formed NOx. The reduction of NOx is favorable in the areas of the flame with high concentrations of NOx and local rich conditions and moderate temperatures. Small hydrocarbon radicals formed during fuel oxidation react with NO leading primarily to the formation of molecular nitrogen and HCN.

BURNER AND FIREBOX DESIGN PRINCIPLES

In a modern ultra-low NOx firebox design, the objective of the designer is to minimize NOx by influencing all three occurring mechanisms. The burner is the prime source of NOx. The combustion characteristics of the burner are very much determining the final NOx emission values of the furnace. Nevertheless, if the combination of burner characteristics and firebox fluid dynamics does not fit, even when applying ultra-low NOx burners, final NOx figures can be two or three folds the figures obtained in a burner test performed in a test-furnace.

Technip has designed modern ethylene furnaces with ethylene capacities of over 200 Thousand Metric Tons per Annum in each firebox cell. The firing required in such a firebox cell is 130 MW or more. To put this heat in, a large portion, typically 50 to 80%, of the heat needs to be supplied by large duty diffusion flame bottom burners. Diffusion flame burners in tall boxes may become unstable due to unpredicted re-circulation patterns. When applying techniques to reduce NOx such as operation at lowest possible excess air and the application of multiple level fuel and/or air staging, the tendency of the flames to become unstable increases. For this reason, Technip applies flame stabilization by adding Low NOx flat flame radiant wall or balcony upshot burners (Figure 2).

When applying ultra-low NOx burners in a modern, high capacity ethylene furnace or steam reformer, Technip applies rigorous CFD-NOx studies during the design stage of the project. For this purpose, Technip in combination with the Politecnico di Milano, has developed a CFD-NOx simulator that is able to predict both burner flame characteristics as well as emission values on NOx and CO.

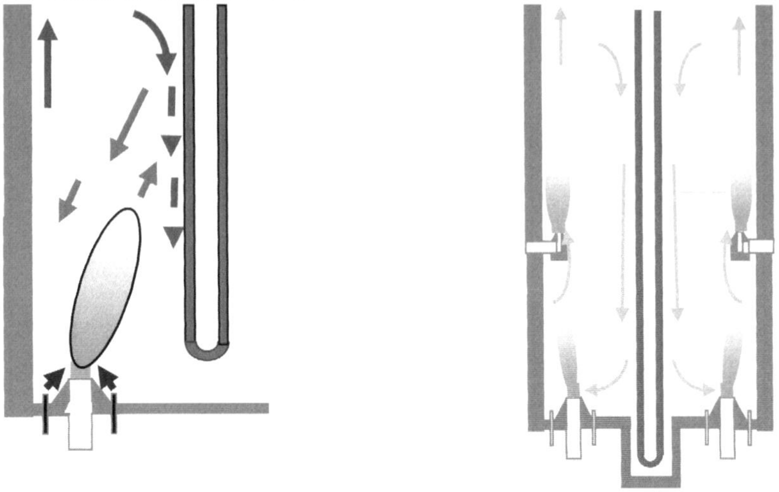

Figure 2a: Flame-Roll Over due to local re-circulation patterns. **Figure 2b: Flame stabilization by Bottom+ Side Wall Combination Firing.**

THE CFD-NOX SIMULATOR

The combustion kinetics model applied in the CFD-NOx simulator is an extraction of the earlier mentioned kinetic scheme and contains some 83 species involved in a reaction network of ~1250 reactions. The reaction network involves, besides the combustion mechanism, the prompt-NOx, thermal NOx and NOx re-reburning mechanisms which accounts for about 260 reactions involving 30 species, amongst others [Frassoldati et al., 2003, Frassoldati et al., 2005]. The mesh of a CFD model for an industrial furnace easily contains more than one million grid cells. To include such a rigorous kinetics model in a CFD simulation leads to excessive and unpractical calculation times. Even with modern high-speed computers equipped with large memory, such an approach is still unfeasible.

However use can be made of the characteristics of the combustion process that allows a different, but feasible approach.

In the combustion process, the overall temperature and fluid dynamics patterns are determined by the combustibles and air fed to the burner and the main flue-gas components. The reactions involved in the formation of pollutants, such as NOx, PAH and unburned hydrocarbons, do not influence significantly the overall temperature, velocity pattern and concentration profiles of the main species.

This phenomenon allows decoupling of the NOx-kinetics from the rigorous CFD simulations [Faravelli et al., 2001]. In the CFD-NOx simulator applied by Technip, first a rigorous CFD model of the furnace is set-up. This CFD model contains a detailed model of one or more burners in a furnace segment.

This CFD model however uses a simplified kinetics model of the combustion process. This kinetics model is sufficiently detailed to allow a proper estimation of the temperature, velocity and concentration profiles of the main flue-gas components.

A network of ideal reactors is defined on the basis of the results obtained by the CFD simulation. The NOx Post Processor, which uses the rigorous combustion model, solves the network of CSTR's for the composition with the given flow and temperature field. After the computations, the results are examined with the CFD package.

NOX POST PROCESSOR

The NOx-Post Processor solves the equations of continuity for the N species (=83) present in the kinetic scheme for the M grid cells (~1.000.000) that are considered to be ideal continues stirred tank reactors (CSTR), as given shown by eq. (1)

$$\nabla \cdot \left(\rho \mathbf{v} x_{i,j} \right) = -\nabla \cdot \mathbf{J}_{i,j} + R_{i,j}\left(C_{i,j}, P_j, T_j, \sigma_j \right) \qquad i = 1, N \; j = 1, M \tag{1}$$

where $\mathbf{J}$ is the diffusion flux (kg·m^{-2}·s^{-1}) and R_i (kg·m-^{3}·s^{-1}) is the net rate of formation or of disappearance of the ith species. The density and the velocity vector are represented by respectively ρ and $\mathbf{v}$. The symbols x and C represent the composition of the fluid in respectively mass fraction and concentration (mol·m-3). The mean temperature and pressure in the CSTR are denoted by T (K) and P (Pa). The turbulent flows $\mathbf{J}_i$ are derived from Fick's law:

$$\mathbf{J}_i = -\left(\frac{\mu}{Sc} \right) \nabla x_i \tag{2}$$

were $Sc = \mu / \rho D$ and the viscosity, μ, (kg·m^{-1}·s^{-1}) and diffusion coefficients, D, (m^{2}·s^{-1}) are effective properties including the turbulent values. One issue that needs to be addressed is the influence of the temperature fluctuations on the kinetic constants and the concentrations, in the evaluation of the net production rates. For this purpose an estimation is made by the CFD model of the average temperature fluctuations inside a grid cell ((σ_j in eq. (1)) and usage is made of this value to correct the kinetic constants.

Currently the available hardware is suited to solve cases with more than 1.000.000 grid cells in an efficient manner. In order to obtain reliable NOx estimates a lumping procedure has been developed to reduce the number of grid cells. This procedure is to lump cells that are geometrical connected to one another and that do not exceed a predefined temperature difference. This lumping procedure typically reduces the number of grid-cells with a factor of 10 to 100. The NOx Post Processor uses the lumped domain for the rigorous calculations of combustion and pollutant compositions.

MODEL VALIDATION

The model has been validated by comparing results with those obtained under controlled laboratory conditions. For this, the well-known and published Sydney bluff-body burner data are used. This burner produces a diffusion flame from a methane/hydrogen mixed fuel. The model results compare very well with published laboratory data.

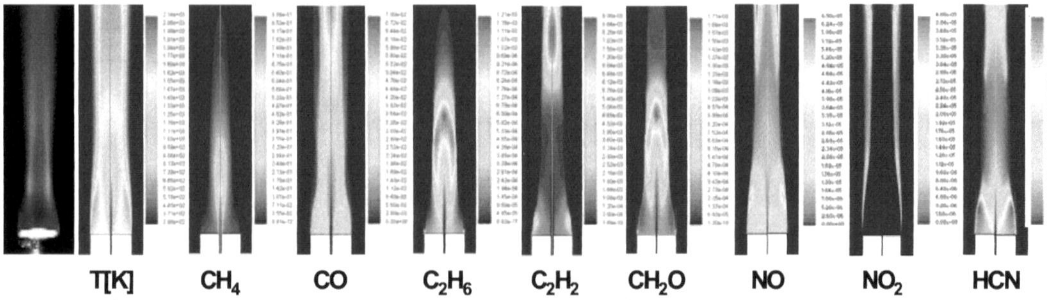

Figure 3: Composition Maps of selected components as predicted by the CFD-NOX Model.

Figure 3 shows the maps of selected components as predicted by the NOx-Postprocessor. NO is significantly produced through the Prompt NOx mechanism. This is confirmed by the considerable intermediate amounts of HCN, which is an active component in the Prompt NOx mechanism. As expected, NO_2 is formed close to the lean side of the flame. In the so-called *neck-zone,* the flame is close to extinction. In this zone, formation of some C_2H_6 can be observed. C_2H_6 is formed through recombination of methyl radicals. Other small amounts of products from hydrocarbon partial oxidation, such as formaldehyde, can be observed as well.

In Figure 4 the comparison between predicted and experimental NO profiles is presented.

Another application example refers to a radiant burner fed with methane and hydrogen/methane mixtures. This burner was investigated in different flameless operating conditions and is placed at Enel Ricerca Laboratories in Livorno [Galletti et al., 2007]. This burner was simulated using a commercial code (Fluent) and the CFD results fed to the NOx post processor to evaluate the pollutants. Details on this simulation can be found in [Frassoldati et al., 2009b]. Figure 5 (left) shows the Temperature and Velocity in the combustor chamber. Moreover, a comparison between measured NO emissions and the corresponding predictions of the NO is also shown in Figure 5.

The agreement is satisfactory and confirms the possibility to use this computational tool also for the simulation of innovative burner and flameless combustion. The validation of the NOx post processor also includes several other laboratory burners [Cuoci et al., 2007], furnaces and combustors for gas turbines [Frassoldati et al., 2009a].

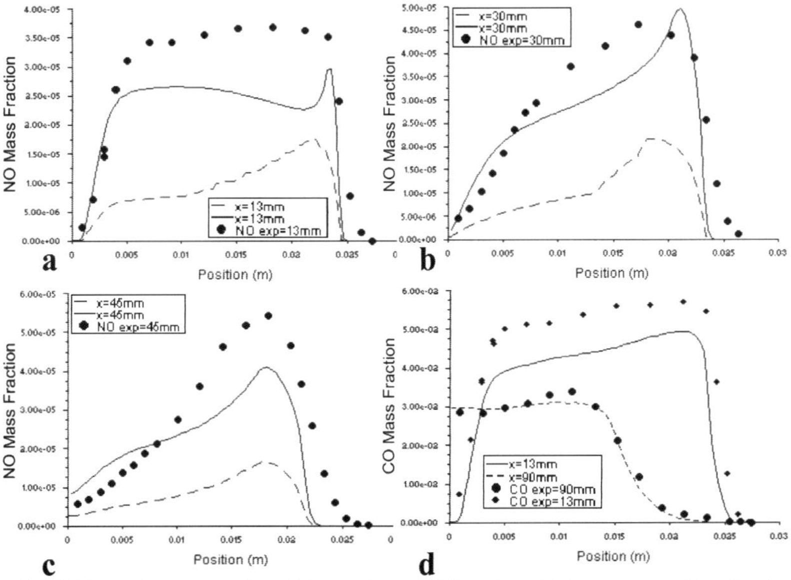

Figure 4: a), b), c) [NO] as a function of radial position at 13, 30 and 45 mm from the bluff body burner calculated using the kinetics Postprocessor. Continuous line: full kinetics mechanism, dashed line: Prompt-NOx suppressed d) [CO] profiles versus radial position at two distances from the bluff body burner calculated using the kinetics Postprocessor. Experimental data from http://www.sandia.gov/TNF/bluffbod.html

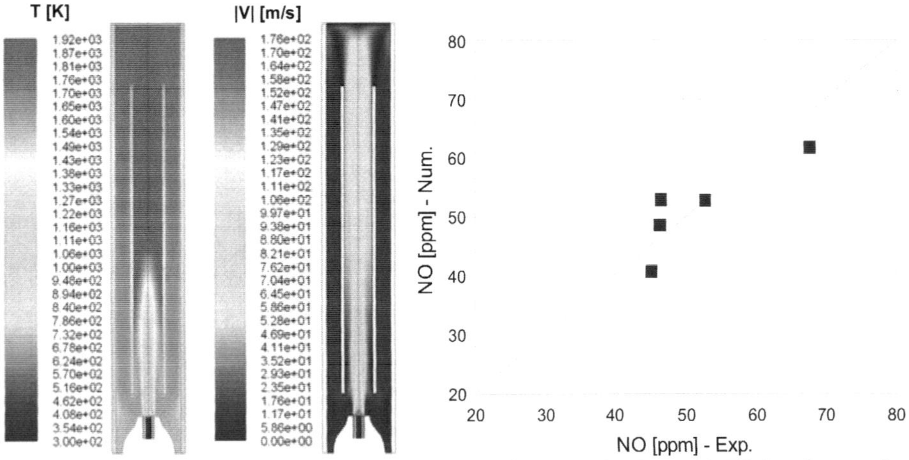

Figure 5 :Left: contour plots of Temperature (K) and velocity magnitude (m/s). Right: Comparison between calculated and measured NO for several cases. Experimental data from [Galletti et al., 2007]

SIMULATION OF ULTRA-LOW NOX BURNER IN AN ETHYLENE FURNACE.

In this case a modern GK6™ ethylene furnace with bottom and side wall firing was studied to predict and optimize the performance of the ultra-low NOx burner. During the design stage, different ultra-low NOx Burner configurations were simulated to study the NOx and flame stability performance. The objective is to find the optimum burner configuration that fulfills the requested heat transfer profile to the tubes, has a stable flame without rollover and meets the requested NOx emission level that is imposed by the pertaining legislation.

In Figure 6, a schematic is shown of the bottom burner segment and the mesh distribution that was used for the CFD+NOx study.

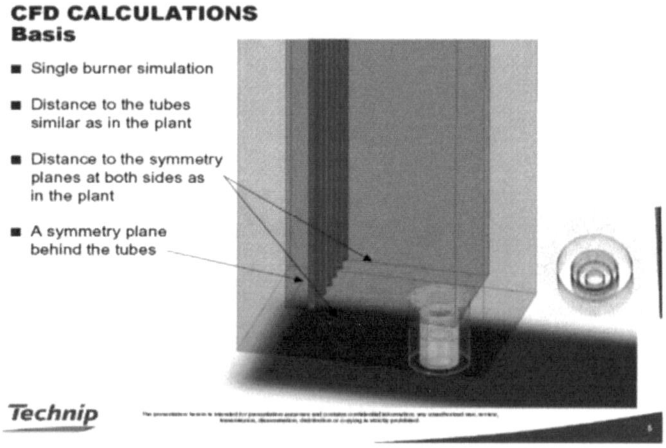

Figure 6: Furnace segment and mesh distribution used for the CFD-NOx study.

In Figure 7, the CFD-NOx results in this ethylene furnace segment of two different burner arrangements are shown showing the temperature distribution in the flame of the ultra-low NOx burner of each of those arrangements. In the first arrangement, significant flame-to-flame interaction occurs which gives a substantial increase in NOx formation (Figure 7a). In the second arrangement, by changing the design of the burner, this interaction is eliminated and NOx substantial reduced. (Figure 7b)

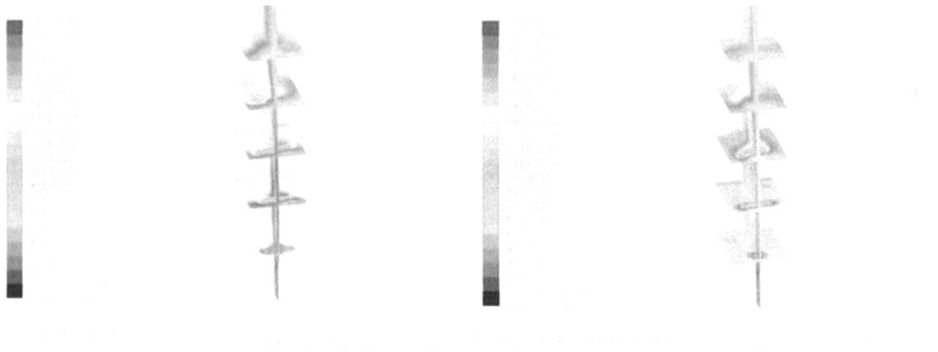

Figure 7:a(left) ,b(right) - Flame Temperature Distribution of two different burner designs.

In Figure 8, the NOx distribution of the two burner designs is shown showing a substantial NOx formation at the burner-burner interaction zone (Figure 7a).

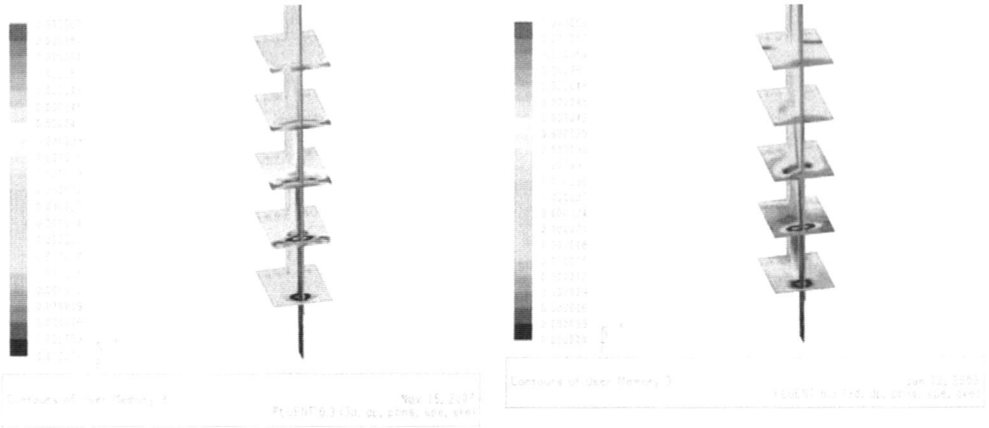

Figure 8: a (left),b(righ) NOx contours at furnace floor and three elevations of two different burner designs. Left figure design showing significant burner-to-burner interaction. Right figure arrangement shows no burner interaction and favorable low NOx formation.

Further investigation of the CFD-NOx simulation results tells that by elimination of the flame-flame interaction the thermal NOx is substantially reduced. An indication of thermal NOx is the formation of O radicals. Figure 9 gives the simulated O radical distribution of the two burner arrangements.

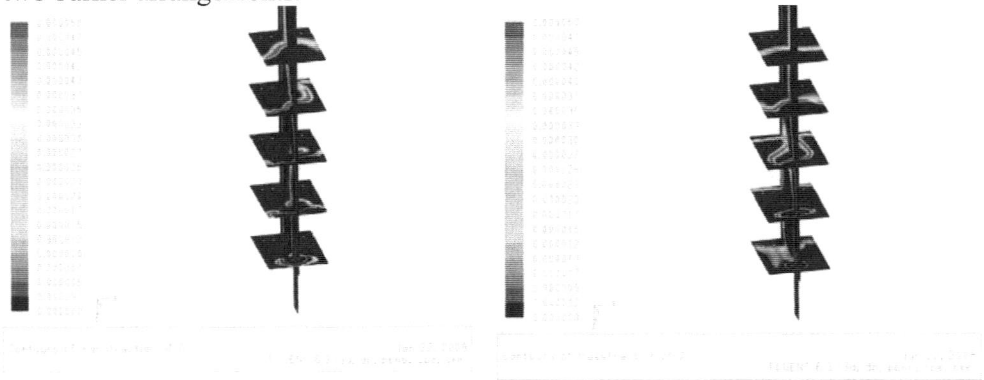

Figure 9a(left),b(right): Oxygen radical distribution in the two burner arrangements.

By changing the burner design also the prompt NOx can be influenced by avoiding local excess hydrocarbons. HCN that is formed as an intermediate component is an indicator of the prompt NOx formation. In Figure 10, the HCN contours in the two burner arrangements are shown showing a significant less pronounced HCN formation in the burner arrangement represented in the right Figure 10b

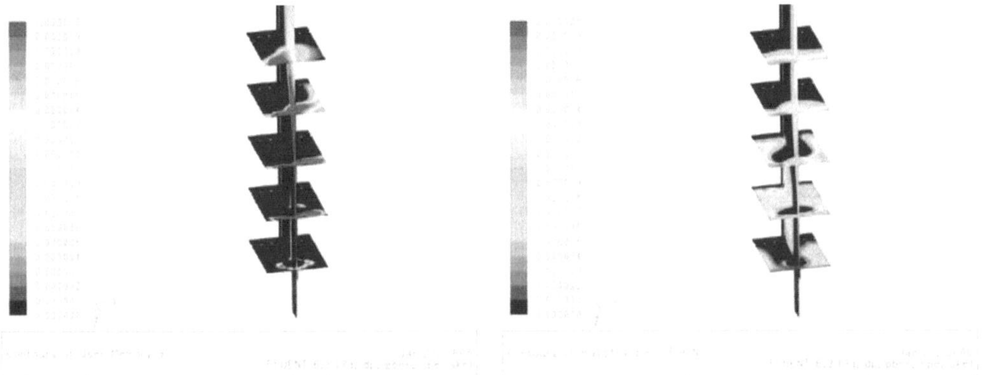

Figure 10: a (left),b(right) HCN Contours as an indicator for Prompt NOx formation for two different burner designs.

CONCLUSIONS

Due to the current legislation in many countries that mandates the reduction of emission levels of pollutant species in industrial furnaces, the design of the burner in combination with the design of the cracking furnace has become a critical factor in ultra-low NOx critical furnace design.

In order to arrive to a sound design, detailed knowledge of the parameters influencing NOx-formation as well as those determining the flame and combustion behavior in the total assembly of burners, radiant coils and firebox, is required.

To obtain such detailed knowledge, Technip, in close cooperation with Scientific Researchers at the Politecnico di Milano, has developed a CFD-NOx simulator that allows accurate prediction of flame behavior and NOx and other emission levels such as CO under industrial conditions. The CFD-NOx simulator has been successfully applied to find the optimum ultra-low NOx Burner arrangement for various steam-reformers and for a GK6™ ethylene furnace that are designed to meet ultra-low NOx emission ceiling levels.

REFERENCES

[Cuoci et al., 2007] Cuoci, A., Frassoldati, A., Ferraris, G., Faravelli, T., and Ranzi, E. (2007). The ignition, combustion and flame structure of carbon monoxide/hydrogen mixtures. note 2: Fluid dynamics and kinetic aspects of syngas combustion. *International Journal of Hydrogen Energy*, 32:3486–3500.

[Faravelli et al., 2001] Faravelli, T., Bua, L., Frassoldati, A., Antifora, A., Tognotti, L., and Ranzi, E. (2001). A new procedure for predicting nox emissions from furnaces. *Computers and Chemical Engineering*, 25:613–618.

[Frassoldati et al., 2005] Frassoldati, A., Colombo, S. F. E., Inzoli, F., and Faravelli, T. (2005). Determination of nox emissions from strong swirling confined flames with an integrated cfd-based procedure. *Chemical Engineering Science*, 60:2851–2869.

[Frassoldati et al., 2009a] Frassoldati, A., Cuoci, A., Faravelli, T., Ranzi, E., Colantuoni, S., Martino, P., and Cinque, G. (2009a). Experimental and modeling study of a low nox combustor for aero-engine turbofan. *Combustion Science and Technology*, 181:483–495.

[Frassoldati et al., 2009b] Frassoldati, A., Cuoci, A., Faravelli, T., Ranzi, E., and Ferraris, G. B. (2009b). Robust and efficient numerical methods for the prediction of pollutants using detailed kinetics and fluid dynamics,. In *19th European Symposium on Computer Aided Process Engineering – ESCAPE19, 14-17 June 2009, Cracow, Poland.*

[Frassoldati et al., 2003] Frassoldati, A., Faravelli, T., and Ranzi, E. (2003). Kinetic modeling of the interaction between no and hydrocarbons at high temperature. *Combustion and Flame*, 135:97–112.

[Galletti et al., 2007] Galletti, C., Parente, A., and Tognotti, L. (2007). Numerical and experimental investigation of a mild combustion burner. *Combustion and Flame*, 151:649–664.

Improving Combustion and Oxidation Reactions: When to use Oxygen

Reed J. Hendershot, Merchant Research and Development, Air Products and Chemicals Inc, Allentown, PA
Timothy D Lebrecht, North America Commercial Technology, Air Products and Chemicals Inc, Allentown, PA
Nancy C Easterbrook, Air Products and Chemicals Inc, Allentown, PA

Today's market demands are everywhere: reduce capital costs, reduce process costs, reduce emissions - especially CO_2, provide more process flexibility, find a way to reduce fuel costs, increase quality and consistency, increase capacity while not increasing costs. The use of oxygen is an excellent way of achieving all of these results. Incorporating oxygen in conjunction with or in place of combustion or reaction air can be used to realize the following results:

- ***Operational savings and fuel savings***: Enriching air with oxygen decreases the overall flow when compared to air. Doing so increases residence time in a heater, reactor, or furnace and minimizes pressure drop in any downstream heat or cooling treatment of outlet gases.
- ***Capital savings***: Oxygen enrichment can be implemented at a fraction of the cost of expanding the original process. Equipment size is decreased by limiting the amount of nitrogen present in the process.
- ***Reduced emissions***: Reduced nitrogen means lower flue gas volumes and thus better flue gas clean up.
- ***Flexibility and reliability***: Operating rates can be changed frequently based upon plant loading. Oxygen enrichment provides the flexibility to constantly match process requirements.
- ***Better quality:*** Oxygen enrichment results in higher combustion temperatures and residence times which contribute to more complete destruction or more complete reactions.
- ***Increased capacity:*** The most common reason for implementing oxygen enrichment is to increase process capacity. Removing nitrogen allows more oxygen to be present enabling greater reaction rates with fewer molecules.

Chemical processes which are amenable to the above list of benefits from the use of oxygen in place of air include oxidations, fermentations, wastewater treatment, and combustion among others. This paper will focus on oxidation and combustion.

Oxygen Enhanced Combustion

Oxygen enhanced combustion (OEC) is used in many different industries including glass, ferrous and nonferrous metals, waste incineration, sulfur recovery units (SRU), fluid catalytic crackers (FCC), and other unit operations (1). New applications for oxygen enhanced combustion are also emerging in biofuels (2), petcoke (3), solid fuels (4), and oxycoal combustion for CO_2 capture (see (5) and other articles in same issue).

Oxygen enhanced combustion can be divided into three operating regimes: low level enrichment, medium level enrichment, and high level enrichment. Low level enrichment is used to define systems where the mole fraction of oxygen in the oxidant stream is between 21 and 28%. This is the simplest and lowest cost implementation since typically oxygen can be added directly to the air main and current air/fuel burners can be used. For higher levels of oxygen enrichment, specialized burners and equipment are needed, but higher levels of benefits are also derived.

Oxygen Enhanced Reactions

Oxygen can also be a key step in the manufacture of a wide variety of industrial chemicals and monomers (6). Table 1 shows a list of major petrochemical oxidation processes that can utilize pure oxygen, oxygen enrichment of air, or simply the oxygen that is present in air.

Table 1. Major Petrochemical Oxidation Products (6).

Chemical	Manufacturing Process
Ethylene Oxide	Oxygen/Air
Propylene Oxide	Oxygen/Air/Chlorine
Acetaldehyde	Oxygen/Air
Vinyl Chloride	Oxygen/Air/Chlorine
Vinyl Acetate	Oxygen
Caprolactam	Oxygen/Air
Terephthalic Acid	Air/Enrichment
Maleic anhydride	Air/Enrichment
Acrylonitrile	Air/Enrichment
Phenol	Air/Enrichment
Acrylic Acid	Air
Acetone	Air
Phthalic anhydride	Air
Isophthalic acid	Air/Enrichment
Acetic anhydride	Air
Formaldehyde	Air
Methyl Methacrylate	Air/cyanohydrins
Adipic acid	Air/nitric acid
1,4 Butanediol	Acetylene/air

Oxygen rather than air leads to improved reaction performance in many cases. It leads to added degrees of freedom and also potential reaction condition versatility. In other words, the use of oxygen can often be justified by improved reaction rates, reaction selectivities, and reaction yields. The reaction of ethylene to ethylene oxide is one such reaction. An article from M. Gans highlights this application (7). Two stages of a three stage process were no longer needed. Nitrogen did not need to be purged after each stage and the use of pure oxygen allowed the reaction to occur at optimum reaction kinetic conditions. The use of oxygen justified itself economically through vastly improved reaction performance and has led to almost universal acceptance of oxygen instead of air (8).

Another reaction where oxygen has an advantage is the production of vinyl chloride monomer – oxychlorination of ethylene. The typical process is based on a fluidized bed catalyst and a reaction system originally developed by Monsanto. Optimum reaction conditions call for an excess of ethylene, and an oxygen concentration which is limited by the lower flammability limit of the system. If air is used, an excess of ethylene cannot be achieved without large ethylene losses. The use of pure oxygen becomes necessary to allow recycle of the desired proportion of reactor gases to achieve optimum reaction conditions.

The use of pure oxygen versus air in chemical reactions must be thoughtfully evaluated. Gunardson highlights several general rules where the use of O2 can be economically justified (6):

- High pressure processes tend to favor the use of oxygen because compression savings offset the cost of oxygen over air.
- Processes with catalysts that have a low conversion per pass favor oxygen because they require recycling unreacted feed and therefore elimination of the inert nitrogen is beneficial.
- Processes that involve toxic or hazardous materials favor oxygen because the vent gas streams are more manageable without the nitrogen acting as a diluent.
- Processes where oxygen is incorporated in the product favor oxygen because oxygen adds value to the product rather than being disposed of in a waste stream.
- Processes that have significant quantities of by-products in the reactor effluent favor oxygen because the by-products can be more readily recovered from a nitrogen free stream.
- Oxidation reactions that are mass transfer limited benefit from high purity oxygen due to the higher partial pressure of reactants without the diluent nitrogen.

Energy Efficiency and Environmental Emissions

From an energy efficiency viewpoint, the nitrogen in combustion air is detrimental. The nitrogen and the small amount of argon make up ~79% on a molar basis of dry air. These gases do not aid in the combustion process but must still be heated up to the same temperature as the combustion products. Since not all of the flue gas enthalpy can be recovered into the process, the exhaust of these gases inherently leads to a net loss on the system. This is displayed graphically using a Sankey diagram for energy use in a furnace in Figure 1. The numbers in the figure correspond to methane combustion in ambient temperature 21% oxygen / 79% nitrogen mixture with a flue temperature of 815 ^{o}C. In Figure 2, the same analysis is displayed graphically for methane combustion in ambient temperature oxygen both with the same flue temperature.

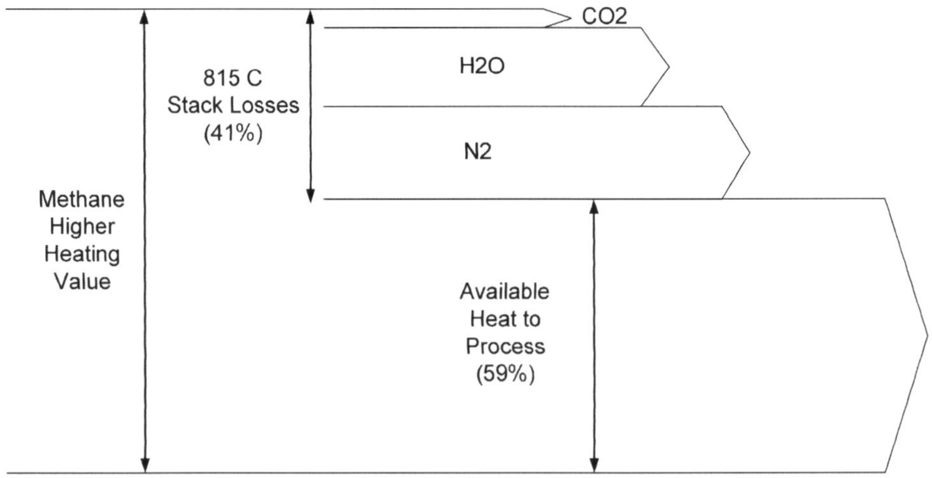

Figure 1: Sankey diagram showing energy distribution for stoichiometric methane and air (21% O2 / 79% N2) combustion in a simplified manner considering only stack losses at 815 C.

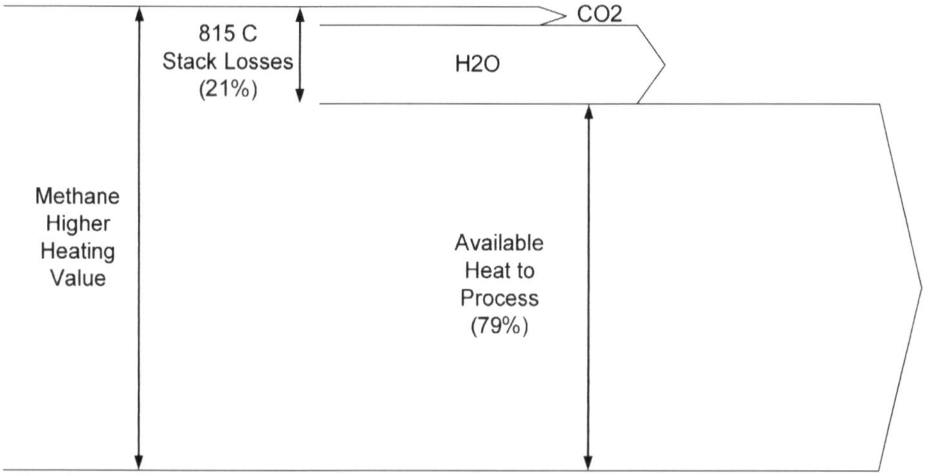

Figure 2: Sankey diagram showing energy distribution for stoichiometric methane and oxygen combustion in a simplified manner considering only stack losses at 815 C.

As the figures show, removing the inert N2 from the combustion air increases the useful heat available to the process from 59% to 79% of the higher heating value leading to an expected fuel savings of 26%. The actual increase in available heat is system dependent but fuel savings on the order of 25-60% are possible using oxygen (1).

As previously mentioned, oxygen also benefits the chemical process of oxychlorination. When using oxygen, the reactor in both the fluid bed and fixed bed configurations is operated at a lower temperature which results in an improved operating efficiency and product yield. The higher heat capacity of the ethylene-rich reaction mixture without nitrogen in the stream has a modulating effect on the operating temperature. Higher operating temperatures are detrimental because they lead to decreased catalyst selectivity and the formation of undesirable chlorinated hydrocarbon by-products. At higher temperatures catalyst activity is decreased and catalyst life is shortened (6). Just as fuel efficiency can be gained in combustion, reaction efficiency can be gained by removing the inert gas nitrogen.

Along with fuel savings, oxy-fuel combustion can also reduce environmental emissions. Decreases in fuel consumption lead directly to reductions in carbon emissions. Since fuel savings on the order of 25-60% can be achieved through the use of oxygen, this leads to the same 25-60% reduction in CO2 emissions at the site. Even when taking into account the energy used in separating the oxygen from air, oxy-fuel and oxygen enhanced combustion can still reduce overall CO2 emissions in many cases. The actual net CO2 reductions will be case dependent because of variability in process fuel, heat recovery, distance to air separation unit, and the carbon intensity of the local power grid.

In addition to carbon emissions, nitrogen oxides (NOx) emissions are strongly influenced by oxygen enhanced combustion. In gaseous fuel systems, thermal NOx by the Zeldovich mechanism (9) is typically the primary source of nitrogen oxides. This reaction depends on both the availability of nitrogen and more importantly the reaction temperature. For air fuel combustion, the limiting factor in NOx production is reaction or flame temperature. For full oxy-fuel combustion, the limiting factor in NOx production is nitrogen availability. There is a range between air and full oxygen combustion in which NOx production increases prior to reductions in NOx when oxygen concentrations of 80-90% are reached (see (1) for more information).

Process and Capital Benefits

The use of oxygen can increase the capacity of many processes with minimal investment in capital. Two example systems where capacity can be increased through the use of oxygen are hydraulically limited systems and heat transfer rate limited systems. In the first case the current equipment does not support increasing the flow rate due to pressure requirements. By replacing some or all of the N2 with O2, some of the hydraulic limitations can be relieved and process flows can be increased. In the second case the heat transfer rate can be increased by increasing the flame temperature through oxygen use. The presence of nitrogen in the combustion process lowers the flame temperature and thereby decreases the radiant intensity of the combustion. The effect of nitrogen on the adiabatic flame temperature is shown in Figure 3.

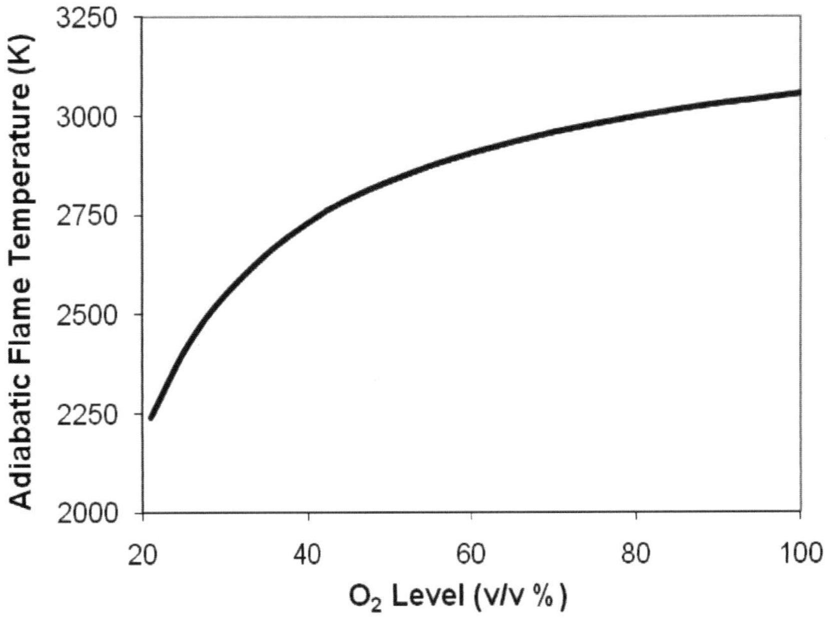

Figure 3: Adiabatic flame temperature for methane combustion in oxygen/nitrogen mixtures.

The effect of higher oxygen concentration in the oxidant cannot be fully described using a thermodynamic analysis of available heat. Since radiation heat transfer is proportional to temperature to the fourth power, an increase in flame temperature with oxygen and changes in flame properties can increase the heat transfer rate over air combustion. Specially designed oxy-fuel burners can take advantage of this by adjusting the flame properties to optimize the radiation wavelength and the radiation properties to maximize efficiency. This has been done with the Air Products Cleanfire ® HR burner (10) (11) for glass melting which has shown increased melting efficiency of 9.2% while using 5% less (12).

Another benefit of oxygen is that it provides flexibility in operations not available with air only operations. For instance, oxygen use can be temporary and employed only when needed. It is possible to increase furnace throughput on a subset of units using oxygen enrichment while another set of units is undergoing modifications or maintenance. In this manner production rates are maintained during partial shutdowns without significant capital investments into spare capacity. Similarly air combustion and oxygen enhanced combustion can be alternately used during a single day. For example in batch furnaces air fuel combustion can be used during furnace hold or charge times while oxygen enhanced combustion can be used during times of high heat load.

Another reaction example of oxygen use is the propylene oxide, isobutene peroxidation process. The process is at 500 to 600 psig, and elimination of nitrogen from the process reduces the gas volume that needs to be compressed. The oxidation reaction has a low conversion of isobutene per pass and elimination of nitrogen from the recycle gas yields savings in compressor size and horsepower. Oxygen is also incorporated in the main product, propylene oxide, and the major by-product, tert-butyl alcohol (TBA) and therefore it has a higher intrinsic value than if it were disposed of as a waste product. These factors combine to make oxygen economically desirable (6).

In addition to the overall process benefits, oxygen use can typically be implemented quickly with a low capital investment. Expanding capacity using air typically requires construction of an additional process line or reaction furnace. In contrast, low level enrichment of air can expand capacity of the exiting process with very minimal cost. Many times the changes can even be implemented while the current process continues to run. Even higher increases in throughput can be achieved by use of higher levels of oxygen.

Case Study

Recently the Ćeská Rafinérská Litvinov facility performed a trial with low level enrichment up to 28% oxygen in a sulphur recovery unit (SRU) using the Claus process. The primary purpose of the trial was to increase the reaction furnace temperature to allow for more complete destruction of ammonia. A secondary purpose of the trial was to test for capacity increase using low level enrichment.

The trial was run by increasing the oxygen concentration in the combustion air by 1-2% at a time and then allowing the furnace conditions to stabilize. Figure 4 shows the air flow and oxygen concentration during the course of the trial while Figure 5 shows the temperature of two different furnace thermocouples during the same time period located at different positions within the furnace. It should be noted that the decrease in furnace temperature toward the end of the run at 22% oxygen is due to a change in the feed composition. Overall, the temperature of the furnace increased by 115 ^{o}C due to an increase in the oxygen concentration of the oxidant from 21% to 28%. This compares remarkably well with a simulation of the process that predicted a temperature increase of 110 ^{o}C.

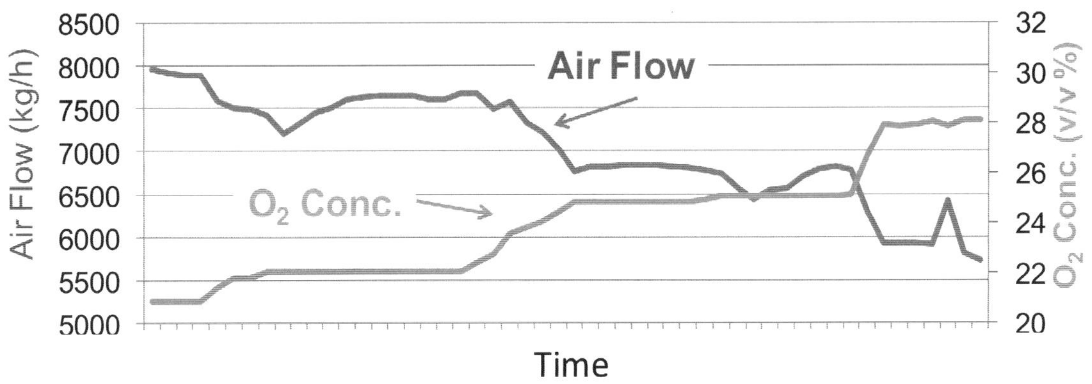

Figure 4: Air flow and oxygen concentration in the oxidant stream during the low level enrichment trial.

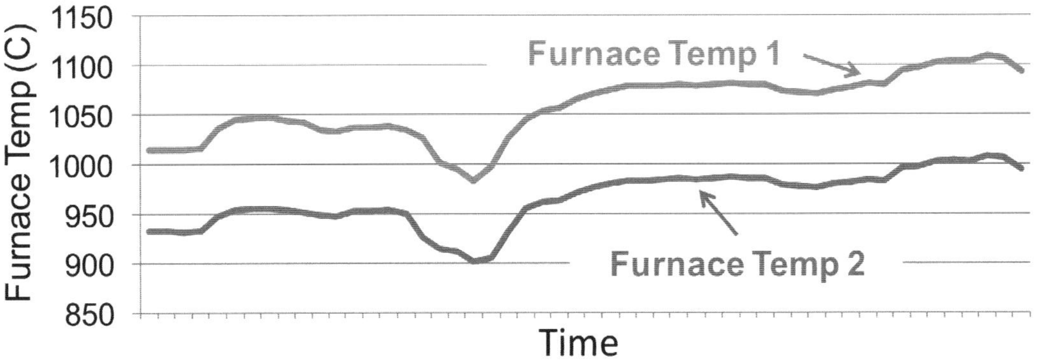

Figure 5: Temperature measurements from two different thermocouples located at different locations within the test furnace.

After increasing the oxygen concentration to 28% and monitoring the furnace temperature, the trial then consisted of testing the capacity increase possible with low level enrichment. Due to a decrease in the air flow to the furnace, the furnace pressure dropped during the test even though the feed acid gas flow increased as shown in Figure 6.

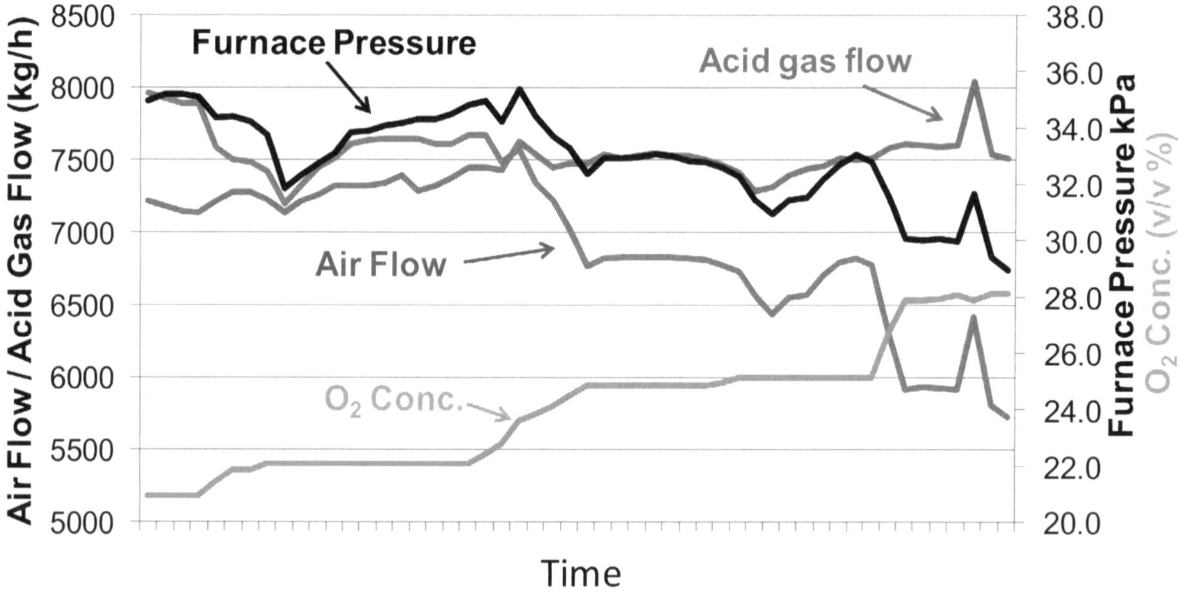

Figure 6: Furnace process conditions during low level enrichment trial.

In Figure 7 additional data are shown from the last increase in acid gas flow near the end of the trial. The final feed gas flow of acid gas of ~8400 kg/hr was 17.6% higher than the baseline conditions at the beginning of the test. Even at this level of feed, the limits of the SRU furnace were not reached but the capacity test was stopped due to a limitation in the availability of acid gas. Therefore the actual capacity increase possible was not demonstrated, but the capacity increase predicted by simulation is 18%.

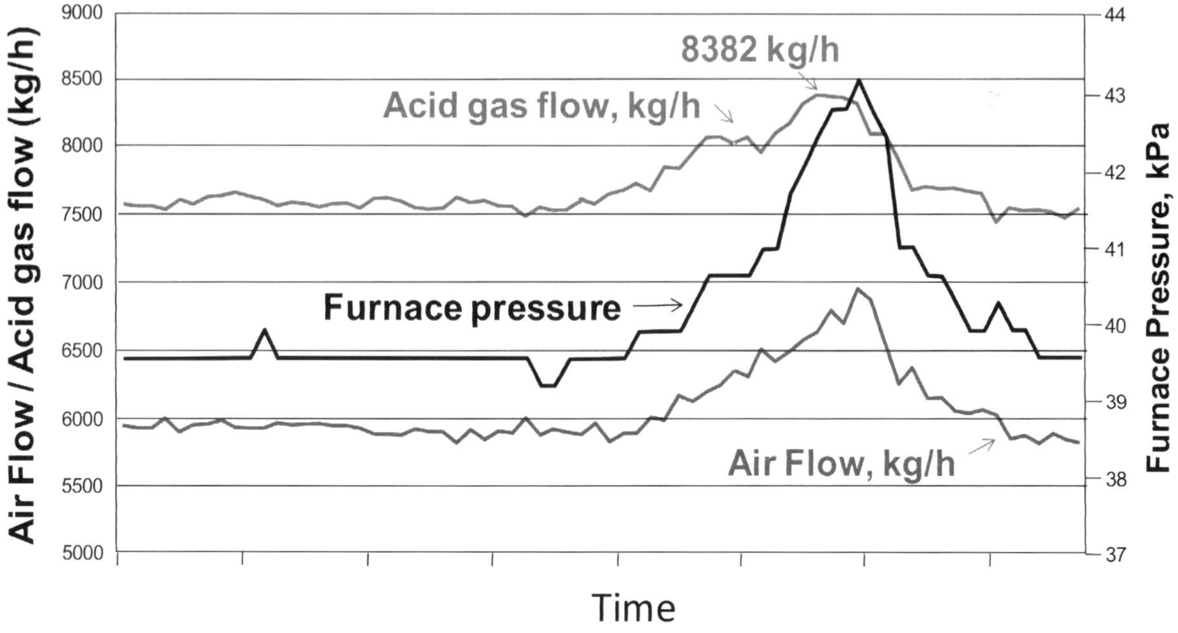

Figure 7: Furnace process conditions during low level enrichment trial at higher time resolution than Figure 6.

Conclusion

Oxygen is a tool which is both powerful and flexible. Engineers and plant managers should evaluate the possibilities of oxygen use within their own processes in order to meet the operational and environmental demands and challenges placed on a chemical facility in these times. However oxygen use requires expert analysis to maximize its benefits for each unique application. Oxygen will help plants achieve operational excellence, reduce costs, achieve environmental compliance through reduced emissions, provide operational flexibility to handle peaks and valleys in product demand or environmental load, increase quality and consistency, and increase capacity, all with minimal capital costs. The use of oxygen is a great way to solve many of today's market and operational challenges.

References

1. Baukal, C E. *Oxygen Enhanced Combustion.* New York : CRC Press, 1998.
2. Johnson, Leighta M, Watson, Matthew J and Slavejkov, Aleksandar G. *Method and Apparatus For Oxy-Fuel Combustion. WO 2008/109482* PCT, September 12, 2008.
3. D'Agostini, Mark D, et al. *Method for Largly Unsupported Combustion of Petroleum Coke. 7,185,595* United States of America, March 6, 2007.
4. D'Agostini, Mark D and Milcetich, Francis A. *Pulverized Solid Fuel Burner. 2008/0184919* United States of America, August 7, 2008.
5. White, Vince, et al., *Purification of oxyfuel-derived CO2.* 2009, Energy Procedia, pp. 399-406.
6. Gunardson, Harold. *Industrial Gases in Petrochemical Processing.* New York : Marcel Dekker, Inc, 1998.
7. Gans, M. Choosing Between Air and Oxygen For Chemical Processes. *CEP.* January 1979, pp. 67-72.
8. Devanney, Michael T. *Chemical Economics Handbook Marketing Research Report Ethylene Oxide,* SRI Consulting, 2007.
9. Zeldovich, Ya B. 1946, Acta Physecochem (USSR), Vol. 21, p. 557.
10. Slavejkov, Aleksandar G, Gosling, Thomas M and Knorr Jr., Robert E. *Method and Device for Low-NOx High Efficiency Heating in High Temperature Furnaces. 5,575,637* United States of America, November 19, 1996.
11. Slavejkov, Aleksandar G, Gosling, Thomas M, and Knorr, Robert E, *Low-NOx Staged Combustion Device for Controlled Radiative Heating in High Temperature Furnaces. 5,611,682* United States of America, March 18, 1997.
12. Tyler, J H, et al., *A Direct Comparison of Oxy-Fuel Burner Technology.* 1999, 59th Conference on Glass Problems. p. 271.

() Status Report: Process Technology Improvements in Supercritical Water Oxidation (C. L.E.A.N. Tech)

Carol A. Blaney, Carol Blaney Consulting, Roswell, GA

This paper will discuss some of the most recent process technology improvements in supercritical water oxidation of wastes, over the last 15 years, and highlight the status of current facilities in operation around the world. Supercritical Water Oxidation Technology established itself about 25 years ago as an emission-free method to destroy toxic wastes, especially aqueous sludges. The technology is very efficient in destroying chlorinated organic toxins such as PCBs and Dioxins that resist destruction in conventional incinerators. As 'green' initiatives have increased, this technology has continually gained momentum because it's efficient and contained, producing pure water, ash, and carbon dioxide by-products.

(Note: A separate session by the same author will focus on the economics of building and running a SCWO facility.)

See more of this Session: Greenhouse Gas Reductions through Industrial Efficiency Improvements
See more of this Group/Topical: Environmental Division

(138f) Measurements of Carbon Dioxide Capture in Various Amines-Ionic Liquids-Water Systems

Yansong Zhao, State Key Laboratory of Multiphase Complex System, Institute of Process Engineering, Chinese Academy of Sciences, Beijing, China

Xiangping Zhang, State Key Laboratory of Multiphase Complex System, Institute of Process Engineering, Chinese Academy of Sciences, Beijing, China

Haifeng Dong, State Key Laboratory of Multiphase Complex System, Institute of Process Engineering, Chinese Academy of Sciences, Beijing, China

Suojiang Zhang, State Key Laboratory of Multiphase Complex System, Institute of Process Engineering, Chinese Academy of Sciences, Beijing, China

With the situation of world wide climate change becoming more and more severe, hunting for the new solvent of high efficiency, low energy consumption and environment-friend to capture carbon dioxide (CO2) from coal industries, petrol industries, and other CO2 emission industries turns to be a crucial task. As known, CO2 could be used in many fields, for instance, refrigeration, farming and synthesis of amines and urea industries. Nowadays, the main and effective method for CO2 capture is based on the chemical absorption and this method could fix much more CO2 quickly. The primary chemical absorbents are monoethanolamine (MEA), methylethanolamine (MDEA), diethonalamine (DEA), and triethonalamine (TEA) or mixtures of them. However, there are some shortcomings of these chemical absorbents, such as corrosion, volatility, toxicity and high energy consumption, which make it necessary to design and synthesize a novel and more effective system to scrub CO2. Ionic liquids, as a new kind of low melting point salt, are focused on by many scientific researchers as their outstanding advantages for CO2 capture: low energy consumption, low pollution level and recyclable properties. However, considering their CO2 absorption abilities, most of ionic liquids could fix less CO2 than that of in amines absorption systems. Recently, in light of CO2 capture properties of amines and ionic liquids, some researchers pay more attentions on the new composite solvents containing amines and ionic liquids. It is reported that these composite systems could fix more CO2 due to the function of amines and be easy to be regenerated because of the existence of ionic liquids and low concentration of water, which means low energy will be used. The conventional ionic liquids, BmimBF4, BmimPF6 and BmimCl, are added into amines to form composite capture systems, and some satisfactory results are obtained in our past work. However, price of these ionic liquids is too high, so cheap and effective ionic liquids, monoethanolamine fluoroborate (MEABF4), N-methyl-diethanolamine fluoroborate (MDEABF4), N-methyl-diethanolamine sulfate (MDEASO4), N-methyl-diethanolamine phosphate (MDEAPO4), and N-methyl-diethanolamine hydrochloride (MDEACl), are synthesized in our lab to form composite absorbents to capture CO2. In this work, amounts of further experiments have been carried out at pressure ranging from 0.00-9.00MPa and temperature between 10°C and 90°C, and some expected results have already been obtained. The main composite solvent systems compounded by amines and ionic liquids are investigated, which performs an effective activity in capturing CO2. 1.000g solvent could absorb as high as 0.158g CO2 in 500 minutes at average by some capture systems. Thermodynamics properties of the various systems are also accurately measured. It is also shown in the research that the captured CO2 could be released very easily from the composite systems, which means less energy will be needed in the regeneration process of absorbents. Besides, CO2 absorption and desorption mechanisms of amines and ionic liquids is also studied in this work. Some new and reasonable mechanisms are discovered according to the characterization results of absorption systems before and after CO2 capture. Keywords: Amines, Ionic Liquids, Carbon Dioxide Capture

See more of this Session: Greenhouse Gas Reductions through Industrial Efficiency Improvements
See more of this Group/Topical: Environmental Division

HYBRID MEMBRANE ABSORPTION PROCESS FOR POST COMBUSTION CO_2 CAPTURE

S. James Zhou and Howard Meyer, Gas Technology Institute, Des Plaines, IL

Ben Bikson and Yong Ding, PoroGen Corporation, Woburn, MA

Abstract

Gas Technology Institute (GTI) and PoroGen Corporation are jointly developing a novel hybrid membrane – absorption process for pre- and post-combustion CO_2 capture (Carbo-Lock™ process). The novel Carbo-Lock™ process combines advantageous features of both absorption and membrane processes to generate cost-effective separation and capture of CO_2 from various emission sources. Together the organizations are making extensive laboratory and bench-scale measurements to bring this technology development to commercialization.

The Carbo-Lock™ process is a hybrid of membrane and the conventional absorption processes. CO_2-containing gas passes through small membrane tubes (hollow fibers with porous walls) while a CO_2 selective solvent (typically an amine solution) flows on the shell side of the membrane tubes. CO_2 passes through the nano-porous membrane and is absorbed in the selective solvent. The CO_2 rich solvent can be regenerated in a second membrane module operated in a reverse process.

The Carbo-Lock™ process uses a novel hollow fiber membrane technology patented by PoroGen. This novel membrane is made from a chemically and thermally stable commercial engineered polymer poly(ether ether ketone) or PEEK. The PEEK membrane contactor can provide a platform for solvent-based systems beyond conventional amines. The reduced size requirements translate to lower solvent inventories, less metal exposure to corrosive liquids, and lower visual impact (smaller footprint) for siting at congested power plants. Since the gas and liquid phases are not directly in contact, the solvent purity levels are expected to remain high and solvent losses, from foaming and carryover, are expected to remain low. Utilizing the membrane contactor in the regeneration section will greatly reduce energy usage for desorbing CO_2 from the solvent. The system operates at very low gas pressure drops, comparable to those achieved with very large diameter columns, and much lower than those of conventional membrane systems.

The preliminary test and computer simulation results of hybrid Carbo-Lock™ membrane-absorption process are presented in this paper.

Introduction

The hybrid membrane/absorption process as shown in Figure 1 is enabled by poly(ether ether ketone), PEEK, based hollow fiber membrane contactor that can replace any column-based mass transfer device in numerous industrial applications such as the column-based devices currently employed in the amine absorption process for CO_2 separation and capture. The membrane is in the hollow fiber configuration and is produced by a patented environmentally benign melt extrusion process. PoroGen is the only manufacturer of the porous PEEK membranes worldwide. The hybrid membrane/absorption device is a contactor that utilizes solvent absorption mechanism with the membrane as the phase boundary between the gas and the liquid. Advanced concepts of our hybrid process employ selective membranes that enhance separation efficiency by functioning as a "selective valve", enabling fast transport of CO_2 while blocking gas components that cause degradation from reaching the absorbent (for example, blocking oxygen from reaching amine) and/or preventing excessive loss of absorbent by evaporation.

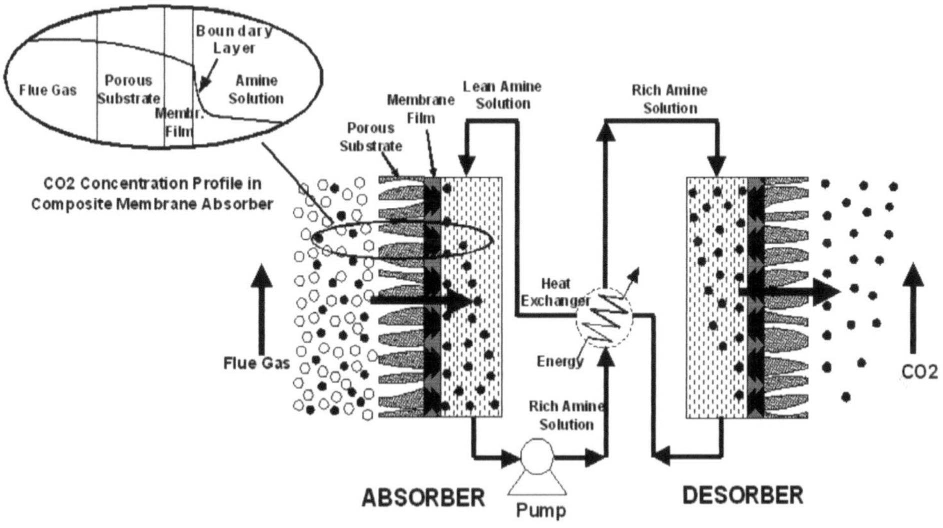

Figure 1 **Integral Membrane Absorption Process for CO2 Capture from Flue Gas**

Process Benefits

The interfacial area is a key variable that directly influences the overall mass transfer and thus the size and performance of the processing equipment. In the PEEK membrane contactor, the interfacial area will be increased by an order of magnitude or greater over conventional packed or tray columns by using super-hydrophobic porous hollow fiber membrane modules. The hollow fiber configuration will allow for high membrane surface area per unit volume and will result in a much more compact absorption system of dramatically reduced size and weight which will also reduce the footprint of the carbon capture island, especially at land-limited sites. For example, the membrane contactor CO_2 removal plant for a 350MW facility is only 11% the size of a conventional packed column based MEA plant.[1] The modular design of membrane contactors will enable the deployment of shop pre-fabricated equipment rather than field fabricated equipment, and will decrease the necessary foundation/footing requirements for structures. Additional benefits include, but are not limited to, the elimination of equipment corrosion due to the inert materials of construction used in hollow fiber contactors (polymeric materials only), decreased solvent losses because there is no carryover nor foaming, lower solvent degradation, and decreased parasitic fan loads because of the decreased pressure and pressure drop required for the hollow fiber membrane contactor vs. conventional membranes and columns.

The membrane is a nanoporous, super-hydrophobic hollow fiber membrane. The novel hollow fiber membrane technology is based on the chemically- and thermally-resistant, commercial engineering polymer, PEEK. The PEEK membrane material utilized is a high temperature engineered plastic that is virtually non-destructible under the operating conditions encountered in typical flue gas and other industrial applications. It can withstand contact with essentially all common treating solvents. This is a major distinguishing feature and major benefit of our technology. The broad benefits of the membrane contactors are well recognized and numerous attempts have been made to deploy them in combination with amine based solvents for CO_2 capture. However, all tested hollow fiber contactors were destroyed or their performance otherwise degraded in contact with the absorption solvent used for CO_2 capture due to the aggressive nature of such solvents. This led to development of specially formulated, less aggressive solvent systems that can be placed into contact with available membranes. However, to date, these specialized systems resulted in inferior CO_2 capture performance.

The Membrane Modules

PGC constructs hollow fiber membrane modules utilizing computer controlled helical winding process. This methodology enables construction of contactor modules with exceptionally uniform and controlled packing density. The structured packing insures optimal mass transfer without channeling or bypassing, low pressure drop and minimizes concentration polarization in liquid boundaries.

Figure 2. Structured packing generated by computer controlled helical winding

Membrane contactor module represents enabling component of this technology. The module consists of the membrane cartridge, o-rings and an external housing. The critical components of membrane module are the hollow fiber membrane and the epoxy tube-sheets. All materials of cartridge construction have been initially tested at high temperature in aggressive solvent environment. To enable efficient counter-current flow, the packing density in the cartridge must be uniform and flow bypassing and entrance and exit effects must be minimized.

Mass Transfer Coefficients

The porous membrane in the membrane contactor typically operates in a non-wetting mode with the liquid side pressure below the breakthrough pressure. Under such conditions given the hydrophobic nature of the membrane and the small pore size, the pores are considered to be completely filled with gas. The gas/liquid interface is thus located at the liquid side pore openings. Depending on the pore size, gas phase mass transfer occurs by molecular diffusion with or without contributions from Knudsen diffusion. The overall mass transfer in the contactor system is controlled by the diffusion rate in the liquid layer.

The mass transfer rate through a membrane contactor is known to be controlled by the liquid side mass transfer due to the 10,000 fold lower CO_2 diffusivity in the liquid phase. Previous data showed that the mass transfer coefficient in the contactor increased with increasing temperature.[2] The mass transfer coefficient is further a strong function of CO_2 partial pressure.[2] In particular, the CO_2 mass transfer coefficient is higher at lower CO_2 partial pressures. It is, therefore, beneficial to operate the membrane contactor at higher temperatures and lower CO_2 partial pressures, conditions that match those of the flue gas. The PEEK membrane can operate at temperatures up to 200°C continuously. This temperature is well in excess of the typical absorption process condition. The high temperature operating capability is of a critical importance in the high temperature regeneration process. To enable CO_2 collection at an elevated pressure in the de-sorption step that will lower the sequestration compression cost, a much higher temperature of regeneration is required.

Initial experiments were carried out to measure mass transfer coefficients as a function of temperature, solvent composition and feed gas flow rates using PEEK hollow fiber contactors. The test module was constructed by computer controlled helical winding and featured structured fiber packing. Representative size modules with 10 ft^2 of membrane area were employed. Synthetic flue gas of a typical flue gas CO_2 concentration and MEA solvent (28%) were utilized. Results are summarized in Table 1.

Table 1. Measured mass transfer coefficients of PEEK hollow fiber contactor

Feed Gas Flow Rate, SCFH	Solvent Rate, L/min	Solvent T, F	Inlet CO_2 Partial P (Kpa)	Kg, mol/(m^2.s.kPa)
25.08	0.74	85	4.11	2.96E-04
25.13	0.755	82.2	3.23	2.93E-04
25.15	0.4	84	3.30	2.74E-04
25.12	0.397	84.4	4.31	2.72E-04
25.16	0.422	124.3	4.71	2.83E-04
25.12	0.41	124.5	3.73	2.83E-04
25.16	0.73	121.7	3.65	2.77E-04
25.06	0.75	115.3	4.51	2.72E-04
34.97	0.585	115.5	5.87	2.62E-04
34.98	0.57	115.5	4.88	2.63E-04
25.16	1.07	122	5.61	4.15E-04
35.1	1.1	120.3	5.68	6.98E-04

These results show that high mass transfer coefficients, Kg, can be attained. The mass transfer coefficients attained are comparable to the best results reported previously.[2,3] Increase in the mass transfer coefficient is expected upon further optimization of hollow fiber morphology and module structured packing design.

Process Economics

The transformational nature of the technology is rooted in its ability to reduce energy and capital costs across a broad spectrum of separation applications. The hybrid membrane/absorption technology combines advantageous aspects of both technologies and cuts across entire spectrum of applications currently served by absorption separations. The technology is a generic technology that enables high mass transfer between gas and liquid streams and can be applied to numerous industrial applications. The technology is compatible with essentially entire spectrum of absorbents from commercial physical and chemical absorbents to experimental advanced absorbents like ionic liquids. In the later case our technology becomes an enabling technology since conventional columns do not function well in contact with viscous ionic liquids. The hybrid membrane/absorption technology carries the promise of dramatic capital and operating cost reduction for CCS.

The process stability and reliability has been one of the key barriers of solvent absorption processes for large-scale CO_2 separation and capture. With the combination of membrane and solvent absorption, the phase boundaries are established by the physical boundaries of the membrane and solvent, making the combined process inherently more stable and reliable than the conventional solvent process.

To estimate the impact that the membrane contactor could have on the carbon capture, GTI used the Integrated Environmental Control Model[4] and Illinois #6 coal for a net 550 MW pulverized coal power plant with constant 2007 dollars as shown in Table 2. This data show that with the conventional column-based MEA process for post-combustion CO_2 separation and capture, the levelized cost of electricity (LCOE) increased by 85.6%; whereas if the membrane contactor based technology is used, the increase is only 42.2% based on membrane module cost of \$100/m^2. The LCOE analysis used the same parameters and assumptions used in reference 5.

Table 2. Comparison of new pulverized coal plants

	Pulverized Coal Boiler		
	PC Subcritical		PC Subcritical with Membrane Contactor for CO_2 Removal
	Case 9[3]	Case 10[3]	
CO_2 Capture	No	Yes	Yes
Gross Power Output (kWe)	583,315	679,923	625,000
Auxiliary Power Requirement (kWe)	32,870	130,310	75,000
Net Power Output (kWe)	550,445	549,613	550,000
LCOE (mills/kWh)[1]	64.0	118.8	91.0[2]

[1] **Based on an 85% capacity factor**

[2] **Value is based on membrane cost of \$100/m^2**

[3] **Case 9 and 10 are from reference 5**

The sensitivity of LCOE to membrane module price is shown in Figure 3. As expected, the LCOE decrease with the decrease in membrane cost.

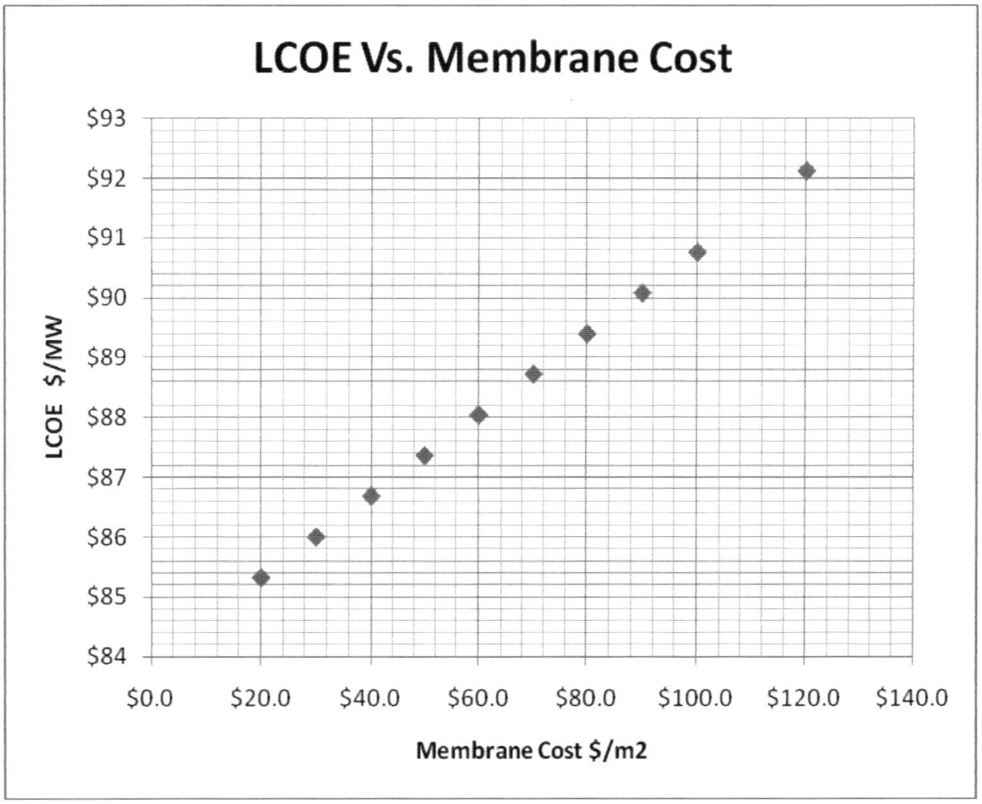

Figure 3. Sensitivity of LCOE to Membrane Price

To reach the goal of 90% CO_2 capture at no more than 35% increase in cost of electricity, the LCOE needs to be less than $87 ($/MW); this represents a membrane cost of less than $40/m^2 as shown in Figure 4. Alternatively, the target can be met by increasing mass transfer coefficient by 150%.

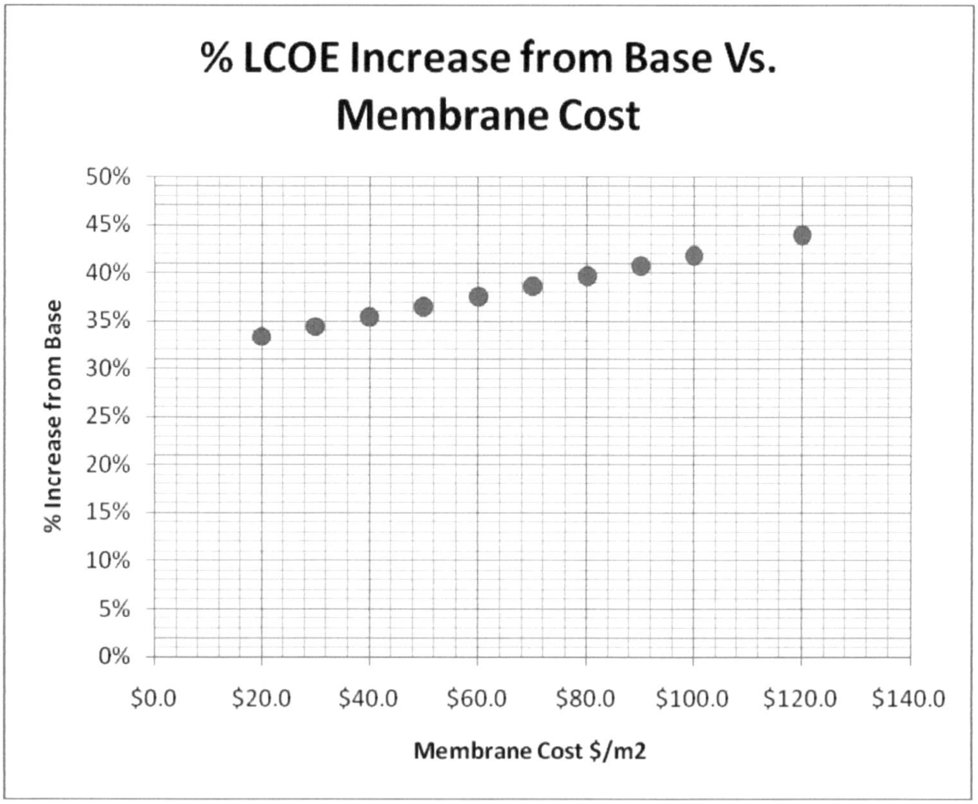

Figure 4. Percent increase in LCOE from PC base Plant(without CO_2 capture) as a function of membrane cost.

Conclusions

Based on the mass transfer measurements and the engineering estimate, the hybrid membrane/absorption process is a promising technology for carbon capture applications with numerous advantages over the current tray or packed column based technology. The new technology carries the promise of attaining the goal set by DOE of the 90% carbon capture at no more than 35% increase in.the cost of electricity generation.

Acknowledgement

GTI gratefully acknowledges the financial support provided by the Illinois Department of Commerce and Economic Opportunity through the Office of Coal Development and the Illinois Clean Coal Institute.

References

1. David C. Thomas, "Carbon Dioxide Capture for Storage in Deep Geologic Formation – Results from the CO_2 Capture Project," Elsevier, 2005, 133-155.
2. Karl A. Hoff, "Modeling and Experimental Study of Carbon Dioxide Absorption in a Membrane Contactor," PhD Thesis, Norwegian University of Science and Technology, March 2003.
3. Shuiping Yan, et. al. "Comparative Analysis of CO_2 separation from flue gas by membrane gas absorption technology and chemical absorption technology in China," Energy Conv. And Manag. **49** (2008) 3188-3197.
4. http://www.iecm-online.com/
5. "Cost and Performance Baseline for Fossil Energy Plants," DOE/NETL-2007/1281.

(142b) Post-Combustion CO_2 Capture Using An Efficient Membrane Process

Xiaotong Wei, Membrane Technology and Research, Inc., Menlo Park, CA

Timothy C. Merkel, Membrane Technology and Research, Inc., Menlo Park, CA

Haiqing Lin, Membrane Technology and Research, Inc., Menlo Park, CA

Bilgen Firat, Membrane Technology and Research, Inc., Menlo Park, CA

Jenny He, Membrane Technology and Research, Inc., Menlo Park, CA

Karl D. Amo, Membrane Technology and Research, Inc., Menlo Park, CA

Richard Baker, Membrane Technology and Research, Inc., Menlo Park, CA

Production of power from coal inevitably generates carbon dioxide (CO_2) as a by-product. To mitigate global climate change, this CO_2 must be captured and sequestered. With current technologies, separating CO_2 from flue gas accounts for about 70% of the total cost for CO_2 capture and sequestration. Among a number of emerging technologies that are being evaluated for CO_2 capture, membrane technology is attractive because of its inherent advantages such as high energy efficiency, a small footprint, environmentally friendly operation (no chemicals), mechanical simplicity, and proven reliability.

Working with NETL, MTR has developed new membranes and process designs to recover CO_2 from power plant flue gas. The membranes have CO_2 fluxes ten times higher than standard commercial membranes, combined with useful CO_2/N_2 selectivities. Based on our best current process design, these membranes offer the potential to capture 90% of the CO_2 in the flue gas using about 12% of the power produced by the plant, at a cost of \$20-\$30/ton of CO_2 captured. Successful commercialization will provide a technology that can be used in the near-term to reliably retrofit existing coal-fired power plants.

Currently, MTR is working with Arizona Public Service (APS) to conduct a field test of this membrane CO_2 capture process at APS' Cholla, Arizona coal-fired power plant. The demonstration will capture 1 ton CO_2/day from real flue gas starting at the beginning of 2010. Lab and field test results to date and future plans will be discussed in this presentation.

See more of this Session: Post-Combustion Techniques for Capture/Separation of Carbon Dioxide II
See more of this Group/Topical: Topical 2: Advanced Fossil Energy Utilization

(142c) Characterization of Coal Flame Impacts Under Oxy-Combustion

Bradley Adams, Reaction Engineering International, Salt Lake City, UT

Andrew Fry, Reaction Engineering International, Salt Lake City, UT

The future use of coal in the United States depends on technologies being made available to capture and store CO2 emissions from power plants. A key candidate CO2 capture technology is oxy-firing of coal. Using oxygen instead of air to combust the coal produces a flue gas stream that it composed primarily of CO2 and water, making subsequent CO2 capture relatively easy. However, application of oxy-firing to existing power plants presents unquantified challenges as the characteristics of oxy-firing compared to air-firing have not been fully determined.

This presentation will draw results from an on-going DOE-sponsored program on oxy-combustion impacts in existing coal-fired boilers. The primary objective of the program is to develop tools to characterize and predict impacts of CO2 flue gas recycle and burner feed design on flame characteristics (burnout, NOx, SOx and fine particle emissions, heat transfer), slagging, fouling and corrosion, inherent in the retrofit of existing coal-fired boilers for oxy-coal combustion. Key elements to be characterized include:

• Appropriate coal feed, oxygen feed, and flue gas recycle (FGR) design for oxy-fired burners or systems;

• Impacts of oxy-firing system design on boiler flame characteristics (burnout, emissions, heat transfer), fouling, slagging and steam tube corrosion relative to the expected behavior in existing air-fired systems.

The proposed research will accomplish this using:

• New experimental data from Sandia National Laboratories' 0.1 kW Entrained Flow Reactor, University of Utah's 100 kW OxyFuel Combustor (OFC), and University of Utah's 1.5 MW L1500 coal-fired furnace;

• Validated mechanisms describing oxy-firing processes;

• Simulation of oxy-combustion in a pilot-scale furnace and a full-scale utility boiler retrofit using oxy-firing design principles.

Information presented here will include background on oxy-combustion flame mechanisms and results from numerical simulations and experiments conducted in a pilot-scale coal-fired furnace that assess flame behavior under air-fired and oxygen-fired conditions.

Three major issues confronted when considering retrofit of existing boilers for oxy-firing are:

• Selection of the flue gas recycle ratio R, here defined as the ratio of the recycle volume to that of the oxygen. Depending on the retrofit objectives, this could be selected to match heat transfer to furnace surfaces rather than the adiabatic flame temperature, of the air-fired furnace.

• Selection of the location from which to recycle the flue gas, that is, selection of the composition of the recycled flue gas based on the location in the flue gas train from which it is recycled back to the furnace.

• Location of oxygen addition and flue gas recycle streams to the boiler.

The selection of R is determined mainly by interest in maintaining the heat flux rate and distribution as close as possible to those achieved with air combustion in order to minimize any design changes to the boiler. Its value depends on the coal, recycle temperature, whether water is condensed or not. It is generally in the range of 2.7 to 3.3 [Kimura et al, 1995; Payne et al., 1989; Wall, 2007; Croiset et al., 2000]. Differences in air and oxy-firing are in large part determined by the composition of the recycled flue gas, which does not depend on R, but does depend on where the recycle stream comes from. Issues related to retrofit of existing boilers to oxy-coal are driven by these differences, which are significant.

Oxy-coal firing provides additional degrees of freedom in the composition of the transport and secondary oxidant streams, namely PO2 in each stream. In order to obtain maximum advantage of the recycle stream on moderating the combustion temperature the recycle stream should be mixed with oxygen as early as possible. Injection early in the furnace is needed to maximize the heat transfer. It is, however, advantage to use some of the CO2 for coal entrainment into the burner and some of the oxygen for flame stabilization. Testing and simulation can quantify the trade-offs for different injection strategies.

The L1500 pilot-scale combustor at the University of Utah is being used to perform experiments detailing firing system design, flame characteristics, together with selected fouling, slagging and corrosion tests. This 5 MBtu/hr (1.5 MW), pulverized coal-fired furnace produces a combustion environment relevant to full-scale utility boilers. In contrast to lab-scale burners, the turbulent length scales in the L1500 are similar to those associated with commercial burners. The furnace is refractory lined, providing realistic boiler temperatures and radiative environment, necessary for heat transfer and corrosion, despite its 10-fold greater surface-to-volume ratio relative to commercial water wall boilers. The furnace has been retrofitted for oxy-combustion with a flue gas recycle system. This modified facility provides maximum flexibility in where and how the FGR, oxygen and coal may be introduced, including staged conditions for reburning of the recycled NOx.

An oxy-research burner has been designed for use in the L1500 pilot-scale furnace. The initial design of the oxy-research burner was based on a patented commercial-scale burner design. This burner design was reduced to a 1.03 MW (3.5 MBtu/hr) firing rate by using a constant velocity scaling technique. The initial design was evaluated and compared to operation of the existing L1500 low-NOx burner through CFD modeling. The modeling results suggested that the constant velocity scaling method did not produce the desired flame behavior for the burner. To correct this behavior, primary gas velocity was reduced, the quarl was lengthened and the burner face setback relative to the quarl was lengthened. These modifications produced the desired flame stabilization location.

The results of this modeling effort suggest that the dominating parameter governing flame shape and stability is not solely velocity. The flame behavior is closely related to particle heating rate, i.e., in order to maintain similar burner behavior when scaling a burner, the particle temperature profile should be maintained. The particle temperature profile is influenced by many factors including: gas velocity, radiation intensity, particle size and hot flue gas recirculation characteristics inside of the quarl. CFD simulations provided an excellent approach to evaluate the combined impacts of all of these factors.

This work will review the burner design and present results from testing of the oxy-research burner in the L1500 furnace under air-fired and oxygen-fired conditions. Results will focus on flame properties (shape, temperature, composition) and heat flux for different recycle gas compositions and oxygen injection strategies.

References

Croiset, E, Thambimuthu KV, Palmer A. Coal Combustion in O2/CO2 Mixtures Compared with Air. The Canadian Journal of Chemical Engineering 2000; 78: 402 - 7.

Kimura, K, Omata K, Kiga T, Takano S, Shikisima S. Characteristics of pulverized coal combustion in O2/CO2 mixtures for CO2 recovery. Energy Conversion and Management 1995; 36: 805-8.

Payne, R, Chen SL, Wolsky AM, Richter WF. CO2 recovery via coal combustion in mixtures of oxygen and recycled flue gas. Combustion Science and Technology 1989; 67: 1 - 16.

Wall, T. F., Combustion Processes for Carbon Capture, Proceedings Combust. Institute, 31 (2007) 31-47.

See more of this Session: Post-Combustion Techniques for Capture/Separation of Carbon Dioxide II
See more of this Group/Topical: Topical 2: Advanced Fossil Energy Utilization

(142d) Equilibrium and Kinetic Studies Carbon Dioxide Capturing On Ca-Based Sorbents

Jorge Pena Lopez, Chemical Engineering Department, University of California at Los Angeles, Los Angeles, CA

Panagiotis (Peter) Smirniotis, Chemical & Materials Engineering Department, University of Cincinnati, Cincinnati, OH

Vasilios I. Manousiouthakis, Chemical Engineering Department, University of California at Los Angeles, Los Angeles, CA

The increase of atmospheric carbon dioxide concentrations in recent years is well documented. Its potential link to an earth wide greenhouse-like effect has led recently to tighter regulations on carbon dioxide emissions from various industrial sources. One of the most common emitters is the fossil fuel based energy production industry which generates high temperature streams containing carbon dioxide at the exit of fossil fuel combustion processes.

We consider the removal of carbon dioxide from high temperature water-air mixtures using calcium-based sorbents. It has been experimentally shown in our prior work that CaO/Zr nanostructured sorbents possess a consistently high capacity to capture carbon dioxide over several high temperature carbonation/decarbonation cycles. In this work, we provide theoretical support for these experimental results by carrying out both equilibrium and kinetic studies. First, we carry out equilibrium calculations by solving to global optimality a Gibbs free energy minimization problem for the multiphase system: N_2, O_2, CO_2, H_2O, CaO, $CaCO_3$, $Ca(OH)_2$. Then, we formulate a simple semibatch reactor model for a thermogravimetric analysis (TGA) apparatus in which the carbonation process is carried out. The predictions of the resulting ordinary differential equation based kinetic model for the CaO adsorption process are compared to TGA obtained experimental data.

See more of this Session: Post-Combustion Techniques for Capture/Separation of Carbon Dioxide II
See more of this Group/Topical: Topical 2: Advanced Fossil Energy Utilization

(142e) Metal-Adeninate Porous Materials for Selective CO2 Capture

Nathaniel L. Rosi, Chemistry, University of Pittsburgh, Pittsburgh, PA

We describe a new class of porous metal-adeninate crystalline materials. We describe the syntheses of these materials and discuss strategies for optimizing the capacity of these materials as selective sorbents for carbon dioxide. In particular, we describe how we can control the coordination chemistry used to construct the materials in order to expose specific Lewis-basic functional groups to the pore cavities in order to facilitate capture of CO2. We also use these materials to systematically evaluate the impact that pore size and volume has on selective CO2 adsorption.

See more of this Session: Post-Combustion Techniques for Capture/Separation of Carbon Dioxide II
See more of this Group/Topical: Topical 2: Advanced Fossil Energy Utilization

Theoretical Screening Good Sorbents for CO_2 Separation: Application to alkali and alkaline-earth oxides and hydroxides

Yuhua Duan
National Energy Technology Laboratory, United States Department of Energy, Pittsburgh, PA 15236, and URS, South Park, PA 15129, USA

Abstract

By combining density functional theory (DFT) and lattice phonon dynamics, the thermodynamic properties of CO_2 absorption/desorption reactions with oxides (M_2O, NO) and hydroxides (MOH, $N(OH)_2$, where M=Li, Na, K; N=Be, Mg, Ca, Sr, Ba) are analyzed. The energy changes and the chemical potential of these reactions have been evaluated. We have analyzed the thermodynamics of the reactions with and without H_2O present. The predicted heats of reactions and the chemical potentials of these reactions versus CO_2 pressures and temperatures are close to the available experimental data once an empirical correction factor is included. We have compared the thermodynamic conditions for these materials with requirements for both pre- and post-combustion capture of CO_2, and found that MgO, $Na_2CO_3/NaHCO_3$, and $K_2CO_3/KHCO_3$ are promising candidates for both pre- and post-combustion capture, whereas Na_2O, and K_2O can form bicarbonates when exposed to CO_2 and H_2O at conditions consistent with post-combustion capture technology at lower operating temperature comparing with CaO. For alkali metal oxides, their CO_2 capture temperatures could be reduced when steam is presented. The results showed that the alkali hydroxides capturing CO_2 to form bicarbonates are less energy cost comparing to forming carbonates. Our results show that $Mg(OH)_2$ are good candidates for post-combustion CO_2 capture, while LiOH and NaOH can be used as CO_2 sorbents both for pre- and post-combustion technologies. Our computational approach can be used as a general scheme for screening a large number of solids for use as CO_2 sorbents.

I. INTRODUCTION

Carbon dioxide is the major product from combustion and is released into the air and causes global climate warming which we are facing today.[1, 2] In order to solve such environmental problem, there is a need to reduce CO_2 emission to atmosphere by capture and sequestration.[3, 4] Current technologies for capturing CO_2 include solvent-based systems (such as Selexol, Rectisol), and alkanolamine-based materials (such as ethanolamine). It is generally accepted that operation or regeneration of these materials is too energy intensive.[1, 2] Hence, there is critical need for new materials that can capture and release CO_2 reversibly with acceptable energy costs. Accordingly, solid sorbent materials have been proposed for capture of CO_2 through a reversible chemical transformation. Among them, re-generable solid sorbents play an important role for CO_2 capture, especially at above room temperature.[1, 5-11]

The solid sorbents containing alkali and alkaline earth metals have been reported in many literatures.[1, 11-19] Among them, their oxides and hydroxides attract more attention from researchers because of their high CO_2 absorption weight percentage and moderate working temperature.[11, 13, 20, 21] Experimental evidences also showed that mixtures of hydroxides with oxides could improve the CO_2 absorption performance.[11, 13]

Comparing with extensive experimental studies of capturing CO_2 with alkali and alkaline earth oxides and hydroxides,[11, 13, 20, 21] the systematical theoretical investigations on their CO_2

capturing properties are very limited. Based on thermodynamic data, Feng *et al*[22] analyzed 11 simple metal oxides and concluded that CaO is thermodynamically the best candidate among them for CO_2 capture in zero emission power generation systems. Based on density functional theory (DFT), Jensen *et al*[23] investigated the CO_2 adsorption on CaO and MgO. Their results showed that CO_2 adsorbs as monodenate on edge sites and bidentate on corner sites on MgO, while on CaO, CO_2 as monodenate on both edge and corner sites. Through *ab initio* evolutionary simulations and high-pressure experiments, Oganov *et al*[24] investigated the high-pressure structures of $MgCO_3$, $CaCO_3$ and CO_2 and their roles in Earth's lower mantle. Recently, combining DFT with lattice phonon dynamics, we developed a novel methodology to select good solid candidates of CO_2 solid sorbents.[15, 25] For solid sorbent Li_2O, we concluded that although pure Li_2O can absorb CO_2 efficiently, it is not a good solid sorbent for CO_2 capture because the reverse reaction, corresponding to Li_2CO_3 releasing CO_2, can only occur at very low CO_2 pressure and/or at very high temperature when Li_2CO_3 is in liquid phase.[15] This predicted results are in very good agreement with experimental measurements.

In this study, we apply our methodology into alkali and alkaline earth metal oxides and hydroxides to explore their CO_2 capture properties systematically, and compare our results with their available thermo-dynamical data.

II. THEORETICAL METHODS

For the CO_2 capture reaction by oxide or hydroxide solids

$$\text{Solid\_A} + CO_2 + [H_2O] \leftrightarrow \text{Solid\_B} + [\text{Solid\_C}] + [H_2O] \quad (a)$$

where [...] are optional, we can ignore the PV contribution and assume that the difference between the chemical potential ($\Delta\mu^o$) of solid phases of A, B(and C) can be approximated by the difference in their electronic energy (ΔE^o) and the vibrational free energy of the phonons.[26] For CO_2, we can treat it as ideal gas. Therefore, the variation of the chemical potential ($\Delta\mu$) for this reaction with temperature and pressure can be defined as[15, 25, 27-29]

$$\Delta\mu(T,P) = \Delta\mu^0(T) - RT\ln\frac{P_{CO_2}}{P_{H_2O}^{\pm 1}} \quad (1)$$

with

$$\Delta\mu^0(T) \approx \Delta E^{DFT} + \Delta E_{ZP} + \Delta F^{PH}(T) - G_{CO_2}(T) \pm G_{H_2O}(T) \quad (2)$$

where ΔE_{ZP} is the zero point energy difference between the reactants and products and can be obtained directly from phonon calculations using following Eq.(6). If reaction (a) doesn't involve H_2O, the pressure of H_2O in Eq.(1) (P_{H_2O}) should set to 1. If the H_2O is a product in reaction (a), plus "+" sign should be taken in Eqs. (1) and (2), and if it is as a reactant, the minus "-" should be used. The free energy of CO_2 (G_{CO_2}) and H_2O (G_{H_2O}) can be obtained by standard statistical mechanics[30]

$$G_{CO_2}(T) \approx \frac{7}{2}RT + \sum_{i=1}^{4}\frac{N_a h\nu_i}{e^{h\nu_i/kT}-1} - TS_{CO_2}(T) \quad (3)$$

$$G_{H_2O}(T) \approx 4RT + \sum_{i=1}^{3}\frac{N_a h\nu_i}{e^{h\nu_i/kT}-1} - TS_{H_2O}(T) \quad (4)$$

where N_a is the Avogadro constant. The entropy of CO_2($S_{CO_2}(T)$) and H_2O($S_{H_2O}(T)$) can be accurately evaluated using the Shomate equation.[31] The vibrational frequencies (ν_i) of CO_2 molecule are 673cm^{-1}(π_u), 1354cm^{-1}(σ_g^+), and 2397cm^{-1}(σ_u^+),[32] and the vibrational frequencies (ν_i) of H_2O molecule are 3657.05cm^{-1}(ν_1), 1594.75cm^{-1}(ν_2), and 3755.93cm^{-1}(ν_3),[33] from which the obtained zero point energies for CO_2 and H_2O molecules are 0.3160eV and 0.5584eV respectively.

In Eq.(2), ΔE^{DFT} is the total energy change of the reactants and products calculated by density functional theory(DFT). In this work, the Vienna *Ab-initio* Simulation Package (VASP)[34, 35] was employed to calculate the electronic structures of solid oxide, hydroxide and carbonate materials. In this study, all calculations have been done using the PAW pseudo-potentials and the PW91 exchange-correlation functional. This computational level was shown to provide an accurate description of oxide systems.[28] Plane wave basis sets were used with a kinetic energy cutoff of 500eV and an augmentation charge cutoff of 605.4eV. The k-point sampling grids of $n_1 \times n_2 \times n_3$, obtained using the Monkhorst-Pack method,[36] were used for these bulk calculations, where n_1, n_2, and n_3 were determined consistent to a spacing of about 0.028Å^{-1} along the axes of the reciprocal unit cells. During calculations, all atoms in the cell were relaxed to the equilibrium configurations. The energies of CO_2 and H_2O molecules are -22.99409eV and -14.27267eV respectively as determined from calculations of an isolated molecule in a cubic box with the length of 20 Å.

Based on phonon calculations under the harmonic approximation, the phonon free energy ΔF^{PH} change between reactant and product solids can be calculated through the free energy of solids described by the Helmholtz form F_{harm} as shown in Eq.(2) and the entropy of the solids (S_{harm})

$$\Delta F^{PH}(T) = \Delta F_{harm}(T) - T\Delta S_{harm}(T) \qquad (5)$$

where the F_{harm}, S_{harm} and the internal energy(E_{tot}) of the solids are defined as [26]

$$F_{harm}(T) = rk_B T\int_0^{\infty} g(\omega)\ln[2\sinh(\frac{\hbar\omega}{2k_B T})]d\omega \qquad (6)$$

$$S_{harm}(T) = rk_B\int_0^{\infty} g(\omega)\{(\hbar\frac{\hbar\omega}{2k_B T})[\coth(\frac{\hbar\omega}{2k_B T})-1] - \ln[1-\exp(-\frac{\hbar\omega}{k_B T})]\}d\omega \qquad (7)$$

$$E_{tot}(T) = \frac{1}{2}r\int_0^{\infty} g(\omega)(\hbar\omega)\coth(\frac{\hbar\omega}{2k_B T})d\omega \qquad (8)$$

where r is the number of degree of freedom in the primitive unit cell. It can be seen that the zero-point-energy (E_{ZP}) can be obtained from Eq.(5) by taking $T \rightarrow 0$.

$$E_{ZP} = \lim_{T\rightarrow 0}(E_{tot}(T)) \qquad (9)$$

where ω is the phonon dispersion frequency, and g(ω) is the phonon density of state. In this study, we employed the PHONON software package[37] in which the direct method is applied following the formalism derived by Parlinski *et al.*[38] In phonon calculations, a 2×2×2 supercell was created from the optimized unit cell structure that was calculated based on DFT. Structures with a displacement of 0.03Å of non-equivalent atoms were generated from the 2×2×2 supercell and DFT calculations were further performed to obtain the force on each atom due to the displacements. These forces were input into the PHONON package[37] to fit the force matrix and further carry out the phonon dispersions and densities, which can be inputted into Eqs.(6)-(8) to calculate the thermodynamic properties.

In the band structure calculations and phonon dispersion calculations, the symbols and coordinates of the high symmetry points in the first Brillouin zone of the crystals are taken from Bradley and Cracknell's definitions.[39]

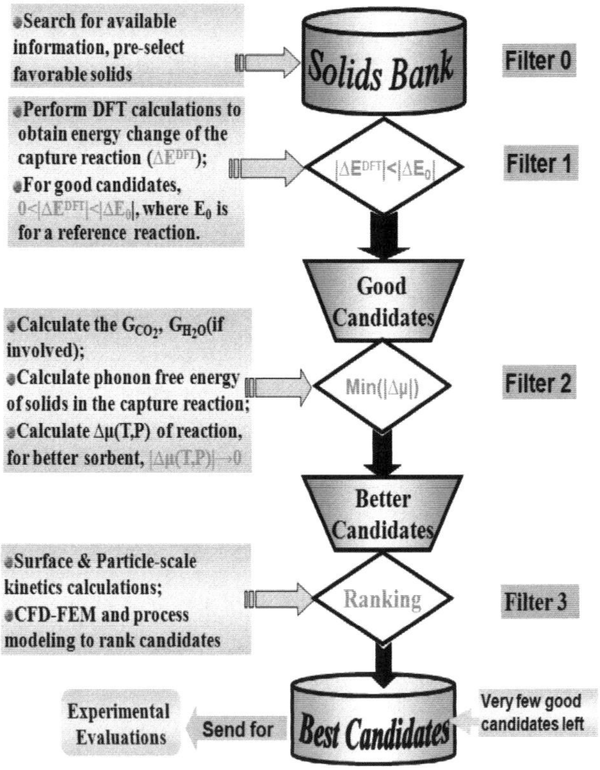

Fig.1 Screening methodology with several filters.

The enthalpy change for the reaction, $\Delta H^{cal}(T)$, can be derived from above equations as

$$\Delta H^{cal}(T) = \Delta\mu^0(T) + T(n_1 S_{CO_2} + n_2 S_{H_2O} + \Delta S_{harm}) \qquad (10)$$

We found that the calculated enthalpy and chemical potential are always higher than the experimental or tabulated data by 20~30 kJ/mol. This is likely due to limitations of DFT. We have subtracted a constant value, ΔH_0, from Eq.(2) in order to empirically correct for this apparent bias. We have found that using ΔH_0 = 26.2 kJ/mol, which is the sublimation enthalpy of CO_2, gives reasonable performance for all systems studied here. We initially thought that ΔH_0 was due to a mismatch in the thermodynamic reference states of the Shomate equation and the DFT total energy calculations. However, this is not the case. Nevertheless, the heat of sublimation of CO_2 does provide a practical estimate of the value of the empirical correction factor.

Figure 1 is illustrated our screening methodology. Generally, for a given data bank, we can manipulate our methodology as several filters:

(0) for each candidate, search its available thermodynamic data to make initial judgment: if it is a promised one or there is no available thermodynamic data, continue to next filter. Otherwise, pick another compound;

(i) Performing DFT calculations for both reactants and products of CO_2 capture reaction and obtain the energy difference (ΔE^{DFT}) of the reaction. If 0.5 eV > ΔE^{DFT} > -2 eV, keep this compound and continue to add on free energy of CO_2 and H_2O (G_{CO2}, G_{H2O} as shown in Eqs.(3)-(4)) to obtain the $\Delta E_1(T)$ for the reaction. Otherwise, go back to filter (i);

(ii) Performing phonon calculations for reactants and products to obtain their zero point energy and phonon free energy changes for the CO_2 capture reactions. According to Eqs.(1) and (2) to obtain the chemical potential ($\Delta\mu(T,P)$) of the reaction. For good candidates, their $\Delta\mu(T,P) \rightarrow 0$ at desired working conditions. After this filter, only a short list of compounds left from the given data bank;

Filter (iii) Additional modeling could be performed to rank the remaining short list of better candidates. One is the kinetics of the capture reactions, which could be done by transport and diffusion calculations as well as experimental measurements. Another necessary and doable modeling is the real behavior of solid in the reactor, which can be done by computational fluid dynamics (CFD) based on finite element method (FEM) approach. These simulations are under progressing. Application of these screening filters will ensure that only the most promising candidates will be identified for experimental testing.

This screening methodology provides a path for rapidly evaluating materials for which experimental thermodynamic data are unavailable. One area where this approach could be used to great advantage is in evaluating mixtures and doped materials, where thermodynamic data are not generally available.

III. RESULTS AND DISCUSSIONS

3.1 DFT and Phonon calculated results

The optimized lattice constants and total energies for the 27 alkali and alkaline earth metal oxides, hydroxides, bicarbonate and carbonates considered in this work are presented in Table I, along with experimental structural data. Since there is no crystal structural data for $LiHCO_3$ and $BeCO_3$ available, we take the structures of $NaHCO_3$ and $SrCO_3$ as their initial structures to obtain their optimized structures. From our calculations, the $BeCO_3$ in the structure of $MgCO_3$ resulted in about 0.5eV higher in energy comparing to that in the structure of $SrCO_3$. It should be pointed out that for each compound it may have more than one known phase been investigated, in Table I only the most stable structure with lowest energy is listed. As expected, the agreement between optimized constants and experimental data generally is very good as shown in Table I. The calculated total energy (E^{DFT}) for each compound will be used to estimate the reaction enthalpy of CO_2 capture reaction as described in the next section.

Based on the optimized structures as shown in Table I, a supercell with at least 2x2x2 size will be built to perform phonon calculations. From their phonon dispersion (ω) and phonon density of states ($g(\omega)$), according to Eqs.(6)~(9), the total energy (E_{tot}), phonon free energy (E_{harm}), entropy (S_{harm}) and the zero point energy (E_{zp}) of each compound can be evaluated out with the temperature dependent. The calculated zero-point energy of each compound is given in Table I.

From Table I, one can see that our calculated entropies of these oxides and carbonates are very close to the experimentally measured values. Our calculated electronic properties and phonon dispersion relations (not shown) are also in good agreement with data from the literature for oxides and carbonates.[27,49-52] Therefore, our results can be reliably used to estimate the thermodynamic properties of the reactions of these oxides capturing CO_2 as presented in the following sections.

Table I. Comparison of the experimental and the DFT calculated structural parameters for the compounds in the reactions studied, with all distances in angstroms and angles in degrees. Although calculations are performed on several phases of each compound, only the stable phase, determined by DFT total energy, is listed in the Table. The zero-point energy and entropy calculated from phonon as well as the experimental data are also listed in the Table.

Compound	Space group & Ref.	Structural parameters		Calculated Energy (eV/f.u.)		Entropy (J/mol.K)	
		experimental	calculated	E^{DFT}	E_{ZP}	Phonon (T=300K)	Exp.[a] (T=298.15K)
Li_2O [b]	$Fm\bar{3}m$(No.225)[41]	a=4.62	a=4.631572	-14.43538	0.23628	38.4681	37.610
LiOH	P4/nmmS (No.129)[42]	a=3.549, c=4.334	a=3.55985, c=4.41752	-15.01799	0.41897	37.0148	42.810
Li_2CO_3 [b]	C2/c(No.15)[43]	a=8.35884, b=4.97375, c=6.19377, β=114.789	a=8.405320, b=5.027886, c=6.294445, β=114.1615	-39.55190	0.59912	93.6157	90.169
$LiHCO_3$	P121/c1(No.14)	No experimental data available. Taken same as $NaHCO_3$	a=3.480953. b=8.987688, c=8.099703, β=115.2319	-38.76293	0.80767	60.3772	
Na_2O	$Fm\bar{3}m$(No.225)[44]	a=5.55	a=5.58517	-11.34789	0.12801	76.8426	75.040
NaOH	Cmcm(No.63)[45]	a=3.4013, b=3.3984, c=11.378	a=3.431009, b=3.414088, c=11.39869	-13.72285	0.36924	60.5562	64.430
Na_2CO_3	C12/m1(No.12)[46]	a=9.01029, b=5.23116, c=6.34548, β=96.062	a=8.951803, b=5.335072, c=6.138608, β=102.205	-37.29272	0.49152	122.5343	138.783
$NaHCO_3$	P121/c1(No.14)[47]	a=3.51, b=9.71, c=8.05, β=111.85	a=3.549292, b=9.795588, c=8.125747, β=111.85	-38.01484	0.77735	*74.6295*	101.700
K_2O	$Fm\bar{3}m$(No.225)[48]	a=6.436	a=6.523623	-10.14413	0.08048	112.137	94.100
KOH	P121/m1(No.11)[45]	a=3.957, b=3.995, c=5.742, β=103.93	a=4.010599, b=4.026439, c=5.930733, β=103.4705	-13.38330	0.33355	62.5907	81.250
K_2CO_3	P121/c1(No.14)[43]	a=5.63961, b=9.8312, c=6.83407, β=98.703	a=5.760548, b=9.904778, c=7.181104, β=97.2946	-36.90480	0.45733	160.1208	155.500
$KHCO_3$	P121/a1(No.14)[49]	a=15.1725, b=5.6283, c=3.711, β=104.631	a=15.46219, b=5.733443, c=3.79398, β=105.7123	-37.88604	0.77091	*81.6815*	115.499
BeO	P63mc (No.186)[50]	a=2.6979, c=4.3772, β=120	a=2.710593, c=4.40006, β=120	-14.36657	0.22559	14.7129	13.770
$Be(OH)_2$	P212121(No.19)[51]	a=4.5301, b=4.621, c=7.048	a=4.50825, b=4.650191, c=7.011854	-29.29364	0.90942	39.4651	53.555
$BeCO_3$	Pnma (No.62)	No experimental data available. Taken same as $SrCO_3$	a=6.762928, b=4.851909, c=7.012976	-36.73083	0.57829	50.8681	52.007
MgO	$Fm\bar{3}m$(No.225[52])	a=4.2198	a=4.248879	-12.00759	0.12611	33.2943	26.950

$Mg(OH)_2$	P$\bar{3}$ m1(No.164)[53]	a=3.1425, c=4.7665, β=120	a=3.168163, c=4.749008, β=120	-27.14938	0.77531	65.8659	63.137
$MgCO_3$	R$\bar{3}$ cH(No.167)[54]	a=4.6338, c=15.0192, β=120	a=4.686491, c=15.13795, β=120	-35.96046	0.53235	69.3531	65.090
CaO	Fm$\bar{3}$ m(No.225)[55]	a=4.8152	a=4.819027	-12.98752	0.11088	39.3738	38.10
$Ca(OH)_2$	P$\bar{3}$ m1(No.164)[55]	a=3.582, c=4.904, β=120	a=3.613773, c=4.96788, β=120	-28.33593	0.71140	82.8147	83.400
$CaCO_3$	R$\bar{3}$ cH(No.167)[56]	a=4.991, c=17.068, β=120	a=5.03979, c=17.12672, β=120	-37.61011	0.4841	95.9951	91.710
SrO	Fm$\bar{3}$ m(No.225)[57]	a=5.1326	a=5.1935	-12.19223	0.07903	56.7416	55.580
$Sr(OH)_2$	Pnam(No.62)[58]	a=9.8889, b=6.1202, c=3.9184	a=10.00019, b=6.046296, c=3.947268	-27.72516	0.72379	93.8210	97.069
$SrCO_3$	Pnma(No.62)[59]	a=6.02, b=5.093, c=8.376	a=6.102091, b=5.156528, c=8.490871	-37.33526	0.46027	104.4270	97.200
BaO	Fm$\bar{3}$ m(No.225)[60]	a=5.539	a=5.660430	-11.95582	0.06492	74.1841	72.00
$Ba(OH)_2$	Pnma(No.62)[61]	a=11.03, b=16.56, c=7.11	a=11.15059, b=16.7613, c=7.102235	-27.63975			107.280
$BaCO_3$	Pmcn(No.62)[62]	a=5.3293, b=8.9275, c=6.6076	a=5.370614, b=9.007695, c=6.552103	-37.52245	0.44584	116.7325	112.100
CO_2	P1		a=20 (fixed)	-22.99409	0.31598		213.388
H_2O	P1		a=20 (fixed)	-14.27267	0.55841		188.832

[a] Taken from HSC Chemistry Package. [63]
[b] Taken from Ref. [15]

3.2 Thermodynamic properties of CO_2 capture reactions

As an example, the heats of reactions as a function of temperature for eight oxides are plotted in Fig. 2. Data from the HSC Chemistry Package[48] are also plotted in this figure for comparison. The discontinuities in the experimental curves indicate phase transitions.

From Fig.2, it can be seen that the calculated ΔH values for Li_2O, K_2O, CaO and BaO are very close to the experimental data, with differences of less than ±5kJ/mol.[18,55] Na_2O and SrO sorbents have discrepancies of about 12 kJ/mol. Among these 8 oxides, the BeO has the largest difference of 15 kJ/mol. This is probably due to the fact that we do not have experimental data for the crystal structure of $BeCO_3$. We used the $SrCO_3$–based structure for $BeCO_3$, as shown in Table I. Differences between this structure and the real experimental structure may be the reason for the larger errors.

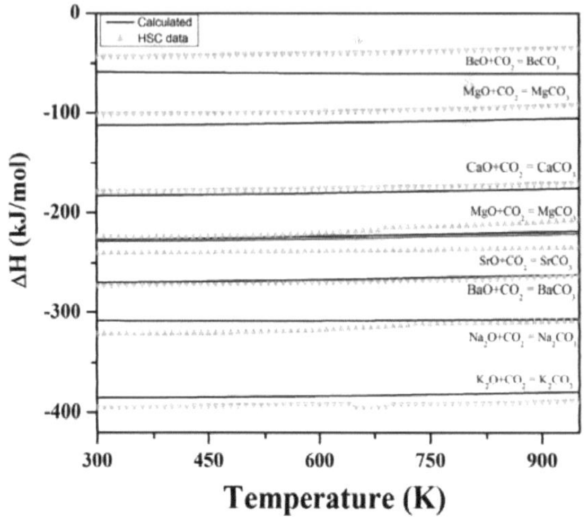

Fig. 2. The calculated heats of reaction as a function of temperature for eight metal oxides. The solid lines were computed from DFT including phonon contributions. The dashed lines were computed from the HSC package. The discontinuities in the HSC curves indicate phase transitions.

As an example, Fig. 3 shows the relationship of calculated chemical potential versus the CO_2

pressure and temperature change. From it, one can see that Overall, by combining DFT with phonon dynamics calculations, we obtained very good predictions comparing with the experimental data. From Fig.3, one can see that for the same reason BeO sorbent has a larger discrepancy at lower temperature range. Our calculated results show that for all of the cases, the DFT-only approach predicted higher CO_2 pressures at each temperature(didn't show in Fig.4) than the DFT+phonon approach.

sorbents are not good candidates for CO_2 capture.

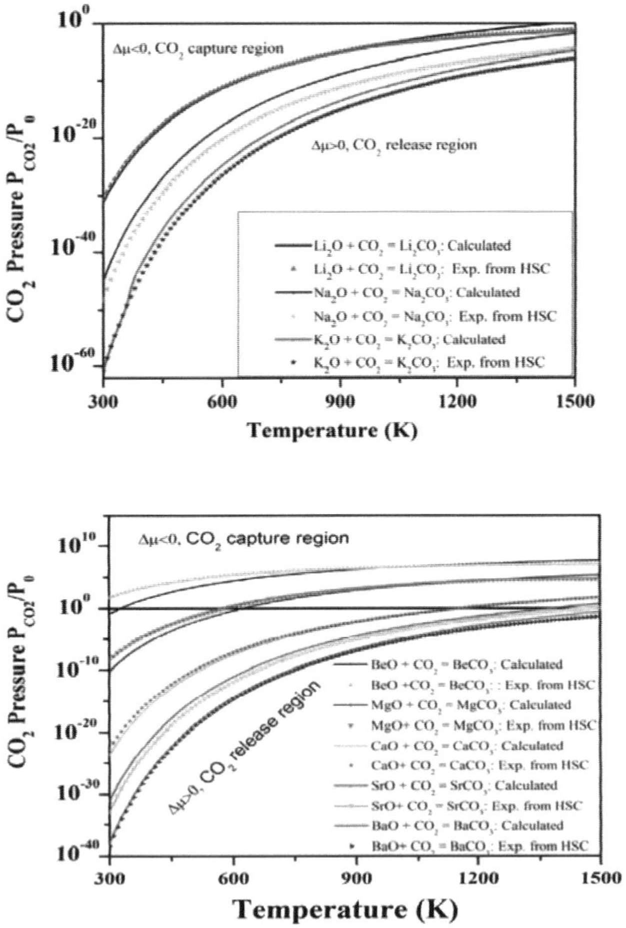

Fig.3. The calculated chemical potentials versus CO_2 pressures (in logarithm scale) and temperatures for the reactions of oxides capturing CO_2 without water involved: (a) alkali metal oxides; (b) alkaline earth metal oxides.

From Fig.3, one can see that up to T=1200K and higher, the Li_2O, Na_2O, K_2O, CaO, SrO and BaO still can absorb CO_2. The CO_2 pressure could be in a big range from very low (~10^{-30} bar) to very high (~10^{10} bar) despite of reaction kinetics, which means during the first half cycle of capturing CO_2, these oxides are very good for CO_2 absorption ($\Delta\mu<0$ region). However, as Fig.3 demonstrated, in the second half cycle of sorbent regeneration to release CO_2, their corresponding carbonates are very stable and only can dissociate ($\Delta\mu>0$ region) at very high temperature. Therefore, these strong

From Fig.3(b), one can see that BeO can absorb CO_2 just above room temperature with lowest pressure of 10^{-2} bar, which is perfect fitting the requirements of the post-combustion capture technology (see section 3.3). Comparing with CaO, MgO is very good for warm temperature CO_2 capture because its turnover temperature is around 600K, which meets the requirements of pre-combustion technology.

Similar results can be obtained for other oxides, hydroxides and carbonates. Table II summarizes our results.

From Table 2, one can see that overall our predicted chemical potentials of the CO_2 capture reactions are quite close to the experimental measurements, which means our theoretical approach can predicted right thermodynamic properties and can be used for investigating unknown systems.

3.3 Applying to pre- and post-combustion CO_2 capture Technology

Under pre-combustion conditions, after water-gas shifting, the gas stream mainly contains CO_2, H_2O and H_2. The partial CO_2 pressure is around 20~25 bar and the temperature is around 300~350°C. To minimize the energy consumption, the ideal sorbents should work at these pressure and temperature ranges to separate CO_2 from H_2. We have estimated the turnover temperatures for pre-combustion conditions of 12 carbonate and bicarbonate reactions. We have omitted reactions involving Li because of the reported instability of $LiHCO_3$ at ambient conditions. This temperature, denoted T_1, is listed in Table III, and is the temperature above which the sorbent will release CO_2. For post-combustion conditions, the gas stream mainly contains CO_2 and N_2, the partial pressure of CO_2 is around 0.1~0.2 bar, and the temperature range is quite different, however, we prefer to the low-temperature (<423K) capture. We also give the turnover temperature for post-combustion capture in Table III, denoted as T_2.

From Table III we see that in the absence of H_2O, the turnover temperatures (T_1, T_2) for Li_2O, Na_2O, K_2O, SrO and BaO are all over 1000K. This means that these oxides are very good at absorbing CO_2, but that their corresponding carbonates are very stable and only can dissociate to release CO_2 at very high temperatures. Therefore, none of these are good sorbents for CO_2 capture under either pre- or post-combustion conditions. CaO could work at high temperature for CO_2 capture under post-combustion conditions (P_{CO2}=0.1 bar) with temperature around 1030 K, which agrees with the experimental results that under P_{CO2}=1 bar, CaO absorbs CO_2 up to 1123 K and is regenerated above 1173 K. [53-54] For these alkali and alkaline earth metal oxides, as shown in Table III, only BeO and MgO can work at lower temperatures for CO_2 capture and release. Although BeO apparently has favorable thermodynamics it is not a viable candidate for CO_2 capture because the powder or dust could cause chronic beryllium disease. Our results show that MgO could be used for both pre- and post-combustion capture technology due to its low regenerating temperature (T_2=560 K for post-combustion condition and T_1=720 K for pre-combustion conditions) which are close to experimental findings.[60]

Table II. The calculated reaction thermodynamics of CO_2 capture by alkali and alkaline-earth oxides.

Reactions	CO_2 wt%	Calculated thermodynamic properties (kJ/mol) ΔH^{cal} and ΔG^{cal} at T=300 K				HSC data[a] at T=300K	
		ΔE^{DFT}	ΔE^{ZP}	ΔH^{cal}	ΔG^{cal}	ΔH(kJ/mol)	ΔG(kJ/mol)
$Li_2O + CO_2 = Li_2CO_3$	142.52	-204.7859	4.5214	-226.731	-179.261	-224.643	-176.290
$Na_2O + CO_2 = Na_2CO_3$	71.01	-284.7066	4.5860	-308.572	-258.100	-322.153	-277.155
$K_2O + CO_2 = K_2CO_3$	46.72	-363.4242	5.8731	-385.508	-335.698	-394.785	-349.084
$BeO + CO_2 = BeCO_3$	129.35	-33.0514	3.5422	-58.514	-5.155	-43.090	1.665
$MgO + CO_2 = MgCO_3$	109.19	-92.5092	8.7089	-112.254	-58.867	-100.891	-48.206
$CaO + CO_2 = CaCO_3$	78.48	-161.7445	5.5229	-182.950	-135.732	-178.166	-130.127
$SrO + CO_2 = SrCO_3$	42.47	-207.3441	6.2967	-228.490	-178.591	-240.494	-188.850
$BaO + CO_2 = BaCO_3$	28.70	-248.2164	6.2658	-269.536	-218.095	-272.491	-220.394
$Li_2O + 2CO_2 + H_2O = 2LiHCO_3$	285.04	-273.0209	19.4379	-285.961	-142.695	-311.043[b]	-189.930[b]
$Na_2O + 2CO_2 + H_2O = 2NaHCO_3$	142.02	-426.5613	22.8017	-436.196	-290.639	-457.494	-311.281
$K_2O + 2CO_2 + H_2O = 2KHCO_3$	93.44	-517.8530	18.2571	-539.023	-382.926	-537.639	-393.799
$Li_2CO_3 + CO_2 + H_2O = 2LiHCO_3$	59.56	-68.2349	13.6847	-85.429	10.366	-86.400[b]	-13.640[b]
$Na_2CO_3 + CO_2 + H_2O = 2NaHCO_3$	41.52	-141.8547	18.2157	-153.824	-58.739	-135.341	-34.126
$K_2CO_3 + CO_2 + H_2O = 2KHCO_3$	31.84	-154.4288	20.2692	-167.922	-51.733	-142.854	-44.716
$2LiOH + CO_2 = Li_2CO_3 + H_2O(g)$	91.88	-76.6589	0.3483	-98.623	-97.134	-94.567	-88.442
$LiOH + CO_2 = LiHCO_3$	183.77	-72.4469	7.0165	-92.026	-43.384		
$2NaOH + CO_2 = Na_2CO_3 + H_2O(g)$	55.02	-108.6058	-0.4371	-134.416	-127.313	-127.510	-122.984
$NaOH + CO_2 = NaHCO_3$	110.03	-158.3597	9.28817	-140.258	-84.497	-131.426	-78.555
$2KOH + CO_2 = K_2CO_3 + H_2O(g)$	39.22	-136.6990	3.1513	-154.110	-157.036	-150.711	-141.079
$KOH + CO_2 = KHCO_3$	78.44	-145.5639	7.76835	-161.018	-114.758	-146.782	-92.897
$Be(OH)_2 + CO_2 = BeCO_3 + H_2O(g)$	102.28	120.0110	-8.65594	88.238	92.326	8.579	16.532
$Mg(OH)_2 + CO_2 = MgCO_3 + H_2O(g)$	75.46	-8.6502	-0.0512	-32.687	-26.177	-19.665	-12.762
$Ca(OH)_2 + CO_2 = CaCO_3 + H_2O(g)$	59.40	-57.9497	1.4598	-80.212	-76.610	-69.035	-64.033
$Sr(OH)_2 + CO_2 + SrCO_3 + H_2O(g)$	36.18	-85.7463	-2.0349	-111.059	-106.685	-105.424	-97.990
$Ba(OH)_2 + CO_2 = BaCO_3 + H_2O(g)$	25.68	-112.0479	-1.24548	-137.121	-131.611	-121.733	-115.686

[a] Calculated by the HSC Chemistry package[48]; [b] from ref. [15]

When H_2O is present, the alkali-metal oxides could capture CO_2 by forming bicarbonates. From Table III one can see that these oxides can be used for post-combustion capture. Comparing with CaO, these oxide sorbents have much higher CO_2 capture weight percentages and can work at lower temperature. Therefore, they are better candidates for CO_2 sorbents.

From Table III, one can see that the $Na_2CO_3/NaHCO_3$and $K_2CO_3/KHCO_3$ are very good candidates for CO_2 sorbents. These materials have favorable turnover temperatures for both of pre- and post-combustion conditions.

From Table III, one can see that NaOH, LiOH and $KLi_2(OH)_3$ can be used for pre-combustion CO_2 capture technology because their turnover T_1 are 880K, 660K and 690K. The turnover T_1 of KOH is much higher with values of 940K and 1290K which are not good for CO_2 capture. As shown in Table III, the turnover T_1 of other hydroxides are too high to be used for pre-combustion CO_2 separation.

Table III. T The highest temperatures (T_1) for sorbents capturing CO_2 at pre-combustion (P_{CO2}=20 bar) and post-combustion (P_{CO2}=0.1 bar) conditions. For reactions involving H_2O P_{H2O}=1 bar.

Reactions	Pre-combustion **T_1 (K)**	Post-combustion **T_2 (K)**
$Li_2O + CO_2 = Li_2CO_3$	hT[a]	1290
$Na_2O + CO_2 = Na_2CO_3$	hT	hT
$K_2O + CO_2 = K_2CO_3$	hT	hT
$BeO + CO_2 = BeCO_3$	370	300
$MgO + CO_2 = MgCO_3$	720	560
$CaO + CO_2 = CaCO_3$	1390	1030
$SrO + CO_2 = SrCO_3$	hT	1230
$BaO + CO_2 = BaCO_3$	hT	hT
$Na_2O + 2CO_2 + H_2O = 2NaHCO_3$	1000	830
$K_2O + 2CO_2 + H_2O = 2KHCO_3$	1120	940
$Li_2CO_3 + CO_2 + H_2O = 2LiHCO_3$	<300	<300
$Na_2CO_3 + CO_2 + H_2O = 2NaHCO_3$	520	450
$K_2CO_3 + CO_2 + H_2O = 2KHCO_3$	460	410
$2LiOH + CO_2 = Li_2CO_3 + H_2O(g)$	hT[a]	hT
$LiOH + CO_2 = LiHCO_3$	660	500
$2NaOH + CO_2 = Na_2CO_3 + H_2O(g)$	hT	hT
$NaOH + CO_2 = NaHCO_3$	880	680
$2KOH + CO_2 = K_2CO_3 + H_2O(g)$	hT	hT
$KOH + CO_2 = KHCO_3$	1290	930
$Be(OH)_2 + CO_2 = BeCO_3 + H_2O(g)$	<300	<300
$Mg(OH)_2 + CO_2 = MgCO_3 + H_2O(g)$	hT	720
$Ca(OH)_2 + CO_2 = CaCO_3 + H_2O(g)$	hT	1450
$Sr(OH)_2 + CO_2 + SrCO_3 + H_2O(g)$	hT	hT
$Ba(OH)_2 + CO_2 = BaCO_3 + H_2O(g)$	hT	hT

[a] hT means the maximum temperature exceeds our temperature range.

IV CONCLUSIONS

In this study, by combining density functional theory with lattice phonon dynamics, a screening methodology was developed and applied to alkali and alkali earth metal oxides, hydroxides, and carbonates. The preliminary results show that our screening methodology can predict the right thermodynamic properties for CO_2 capture reactions and can be used for screening good CO_2 solid sorbents from vast of solid materials which thermodynamic data are unknown. We have demonstrated that first-principles density functional theory can be combined with phonon density of states calculations to provide a reasonable estimate of the thermodynamics of CO_2 capture reactions involving alkali and alkaline earth metal oxides. We have found that an empirical correction factor is needed to bring calculated data into reasonable agreement with data from the HSC Chemistry software package. A single value of this correction factor of about 26 kJ/mol is adequate for all reactions examined in this work. We found that when steam presented the Na_2O and K_2O are better than CaO and could be used for post-combustion CO_2 capture technology, while MgO can be a good sorbent for both post- and pre-combustion CO_2 capture. We have examined the effect of H_2O on the reaction thermodynamics and have found that our modeling approach can be used to account for partial pressures of CO_2 and H_2O and temperatures. We found that formation of bicarbonates from the alkali metal oxides results in a lower sorbent regeneration temperature and that formation of bicarbonate from the carbonates, by addition of CO_2 and H_2O reduces the CO_2 capturing temperature even further. Indeed, we predict that Na_2CO_3 and K_2CO_3 have turnover temperatures for CO_2 capture through bicarbonate formation that are suitable for operation under both pre- and post-combustion conditions.

The alkali metal hydroxides MOH (M=Li, Na, K) can absorb CO_2 to form either bicarbonates or carbonates, but the former has less energy costs and is favorable to be used for CO_2 separation. Among them, $Mg(OH)_2$, NaOH and LiOH are better candidates for CO_2 sorbents. Our calculated thermodynamic properties of these hydroxides are in good agreement with the available thermodynamic data, which indicates that our methodology can produce reliable results and can be used for predicting CO_2 capture behaviors of other solid materials which thermodynamic data is not available.

We have outlined a screening method that will be useful for assessing potential sorbents for CO_2 capture for which no thermodynamic data exist. The methodology involves applying a series of screening filters of increasing computational complexity in order to identify a small number of systems that can be explored experimentally as promising candidates for reversible CO_2 capture.

ACKNOWLEDGEMENT

This work was performed in support of the National Energy Technology Laboratory's Office of Research and Development under contract number DE-AC26-04NT41817. The author thanks Dr. D. C. Sorescu and J. K. Johnson for fruitful discussions.

REFERENCES

[1] D. Aaron and C. Tsouris, Sep. Sci. Technol. **40**, 321 (2005).

[2] C. M. White, B. R. Strazisar, E. J. Granite, J. S. Hoffman, and H. W. Pennline, J. Air Waste Manag. Assoc. **53**, 645 (2003).

[3] E. Ochoa-Fernandez, H. K. Rusten, H. A. Jakobsen, M. Ronning, A. Holmen, and D. Chen, Catal. Today **106**, 41 (2005).

[4] H. Pfeiffer and P. Bosch, Chem. Mater. **17**, 1704 (2005).

[5] J. C. Abanades, G. Grasa, M. Alonso, N. Rodriguez, E. J. Anthony, and L. M. Romeo, Environ. Sci. Technol. **41**, 5523 (2007).

[6] S. C. Lee, H. J. Chae, S. J. Lee, B. Y. Choi, C. K. Yi, J. B. Lee, C. K. Ryu, and J. C. Kim, Environ. Sci. Technol. **42**, 2736 (2008).

[7] S. C. Lee, B. Y. Choi, S. J. Lee, S. Y. Jung, C. K. Ryu, and J. C. Kim, Stud. Surf. Sci. Catal. **153**, 527 (2004).

[8] S. C. Lee, B. Y. Choi, T. J. Lee, C. K. Ryu, Y. S. Soo, and J. C. Kim, Catal. Today **111**, 385 (2006).

[9] V. Manovic and E. J. Anthony, Environ. Sci. Technol. **41**, 1420 (2007).

[10] M. M. Maroto-Valer, Z. Lu, Y. Zhang, and Z. Tang, Waste Manag. **28**, 2320 (2008).

[11] R. Siriwardane, J. Poston, K. Chaudhari, A. Zinn, T. Simonyi, and C. Robinson, Energy Fuels **21**, 1582 (2007).
[12] V. Nikulshina, N. Ayesa, M. E. Galvez, and A. Steinfeld, Chem Eng J **140**, 62 (2008).
[13] S. C. Lee and J. C. Kim, Catal Surv Asia **11**, 171 (2007).
[14] S. C. Lee, H. J. Chae, S. J. Lee, Y. H. Park, C. K. Ryu, C. K. Yi, and J. C. Kim, J Mol Catal B-Enzym **56**, 179 (2009).
[15] Y. Duan and D. C. Sorescu, Phys. Rev. B **79**, 014301 (2009).
[16] K. Essaki, K. Nakagawa, M. Kato, and H. Uemoto, J. Chem. Eng. Jpn **37**, 772 (2004).
[17] J. Ida and Y. S. Lin, Environ. Sci. Technol. **37**, 1999 (2003).
[18] H. Pfeiffer, C. Vazquez, V. H. Lara, and P. Bosch, Chem Mater **19**, 922 (2007).
[19] N. Togashi, T. Okumura, and K. Oh-ishi, J Ceram Soc Jpn **115**, 324 (2007).
[20] J. C. Abanades, E. J. Anthony, J. Wang, and J. E. Oakey, Environ. Sci. Technol. **39**, 2861 (2005).
[21] G. X. Hu, H. Huang, and Y. H. Li, Int. J. Hydrogen Energy **33**, 5422 (2008).
[22] B. Feng, H. An, and E. Tan, Energy Fuels **21**, 426 (2007).
[23] M. B. Jensen, L. G. M. Pettersson, O. Swang, and U. Olsbye, J. Phys. Chem. B **109**, 16774 (2005).
[24] A. R. Oganov, S. Ono, Y. M. Ma, C. W. Glass, and A. Garcia, Earth Planet. Sci. Lett. **273**, 38 (2008).
[25] Y. Duan, in *Proceedings of 7th Annual Conference on Carbon Capture and Sequestration*, Pittsburgh, 2008.
[26] M. Sternik and K. Parlinski, J. Chem. Phys. **123** (2005).
[27] S. Cristol, J. F. Paul, E. Payen, D. Bougeard, S. Clemendot, and F. Hutschka, J. Phys. Chem. B **106**, 5659 (2002).
[28] Y. Duan, Phys. Rev. B **77**, 045332 (2008).
[29] J. H. Wang and M. Liu, J. Power Sources **176**, 23 (2008).
[30] R. G. Mortimer, *Physical Chemistry* (Academic Press, 2000).
[31] M. W. J. Chase, J. Phys. Chem. Ref. Data, **Monograph 9**, 1 (1998).
[32] F. Gygi and G. Galli, Phys. Rev. B **52**, R2229 (1995).
[33] M. Chaplin, Water Structure and Science, http://www.lsbu.ac.uk/water (2008).
[34] G. Kresse and J. Hafner, Phys. Rev. B **47**, 558 (1993).
[35] G. Kresse and J. Furthmuller, Phys. Rev. B **54**, 11169 (1996).
[36] H. J. Monkhorst and J. D. Pack, Phys. Rev. B **13**, 5188 (1976).
[37] K. Parlinski, Software PHONON (2006).
[38] K. Parlinski, Z. Q. Li, and Y. Kawazoe, Phys. Rev. Lett. **78**, 4063 (1997).
[39] C. J. Bradley and A. P. Cracknell, *The mathematical theory of symmetry in solids* (Clarendon press, Oxford, 1972).
[40] S. V. Alapati, J. K. Johnson, and D. S. Sholl, Phys. Chem. Chem. Phys. **9**, 1438 (2007).
[41] T. W. D. Farley, W. Hayes, S. Hull, M. T. Hutchings, and M. Vrtis, J. Phys.: Condens. Matter **3**, 4761 (1991).
[42] S. L. Mair, Acta Crystallogr A **34**, 542 (1978).
[43] Y. Idemoto, J. W. Richardson, N. Koura, S. Kohara, and C. K. Loong, J. Phys. Chem. Solids **59**, 363 (1998).
[44] E. Zintl, A. Harder, and B. Dauth, Z. Anorg. Allg. Chem. **198**, 88 (1931).
[45] H. Jacobs, J. Kockelkorn, and T. Tacke, Z. Anorg. Allg. Chem. **531**, 119 (1985).
[46] I. P. Swaninson, M. T. Dove, and M. J. Harris, J. Phys.: Condens. Matter **7**, 4395 (1995).
[47] B. D. Sharma, Acta Crystallographica **18**, 818 (1965).
[48] P. Touzain, F. Brisse, and M. Caillet, Can. J. Chem. **48**, 3358 (1970).
[49] J. O. Thomas, R. Tellgren, and I. Olovsson, Acta Crystallogr B **B 30**, 2540 (1974).
[50] D. K. Smith, H. W. Newkirk, and J. S. Kahn, J. Electrochem. Soc. **111**, 78 (1964).
[51] R. Stahl, C. Jung, H. D. Lutz, W. Kockelmann, and H. Jacobs, Z. Anorg. Allg. Chem. **624**, 1130 (1998).
[52] P. Karen, A. Kjekshus, Q. Huang, and V. L. Karen, J. Alloys Compd. **282**, 72 (1999).
[53] T. Nagai, T. Hattori, and T. Yamanaka, Am. Mineral. **85**, 760 (2000).
[54] F. A. Bromiley, T. B. Ballaran, F. Langenhorst, and F. Seifert, Am. Mineral. **92**, 829 (2007).
[55] C. H. Shen, R. S. Liu, J. G. Lin, and C. Y. Huang, Mater. Res. Bull. **36**, 1139 (2001).
[56] H. Sitepu, B. H. O'Connor, and D. Li, J. Appl. Crystallogr. **38**, 158 (2005).
[57] O. J. Rutt, G. R. Williams, and S. J. Clarke, Chem. Commun. **2006**, 2869 (2006).
[58] H. W. Grueninger and H. Baernighausen, Z. Anorg. Allg. Chem. **368**, 53 (1969).
[59] W. Pannhorst and J. Loehn, Z. Kristallogr. **131**, 455 (1970).
[60] L. G. Liu and W. A. Bassett, Journal of Geophysical Research **77**, 4934 (1972).
[61] H. Lueke, Acta Crystallogr. B **24**, 1705 (1968).
[62] S. M. Antao and I. Hassan, Phys. Chem. Miner. **34**, 573 (2007).
[63] HSC Chemistry software 6.1, Pori: Outotec Research Oy, www.outotec.com/hsc (2006).

Utilization of Carbon Dioxide towards Control/Suppression of Underground Coal Fires

Ali S. Rangwala,
Assistant Professor
Fire Protection Engineering, Worcester Polytechnic Institute,
Worcester, MA

Evan J. Granite,
National Energy Technology Laboratory,
Pittsburgh, PA

Abstract
Underground coal fires represent a substantial nuisance in terms of property damage, public safety, greenhouse gas generation, and coal resource destruction. Some of the techniques employed to combat these fires will be illustrated. The magnitude of greenhouse gas generation and coal consumption will be estimated. Future research areas into this important issue such as improved fire fighting, detection, and prevention will be discussed. The possible utilization of carbon dioxide towards control and suppression of underground coal fires will be examined.

Introduction

Uncontrolled underground coal fires have been reported in all major coal-producing countries, including: Australia, China, India, Indonesia, Russia, South Africa, and the United States. The fires result in a direct loss of valuable resources through the unwanted burning of coal reserves. For example Heffern and Coates[1] estimate that somewhere between 33 billion and several hundred billion metric tons of coal has burned naturally in the Powder River Basin (North America) over the last four million years, most of which burned within the last two million years. Such fires have been an important source of greenhouse gas emissions in the geological past. There is also an indirect loss of coal resources in underground coal-mine fires because access to coal seams may be blocked. Underground coal fires in coal mines can also depreciate the quality of groundwater (low pH and high concentration of dissolved solids) seeping from the mine [2].

Coal fires usually start at a surface outcrop of a coal seam with a certain thickness that has been exposed by erosion [3]. Several studies have found that the oxidation rate of coal increases dramatically as the temperature crosses a threshold of ~70 °C to 80 °C [3] This temperature can be easily achieved by strong solar heating[4]. Coal fires can also be initiated in coal mines due to spontaneous combustion, which is the slow oxidation process that can occur due to relatively low exothermic reactions and the consequent build up of heat due to the low conductivity environment which does not allow for heat to be dissipated. This slow gradual build up of heat (which can last for days) ultimately builds up enough temperature to initiate oxidative smoldering combustion of the coal. Spontaneous combustion in underground mines is especially dangerous because the atmosphere may be inherently explosive, and a highly localized incident of self-heating can initiate an explosion involving the entire mine. Underground coal mine fires are hard to locate and extinguish mainly because they so deep seated and difficult to reach.

An illustrative sketch representing the hazards associated with a coal fire is shown in figure 1. The illustration is adapted from recent work on water based suppression of underground coal fires published by Hadden and Rein [5]. Fire fighting efforts are shown on the left hand side. The damage to infrastructure—roads, houses, ground collapse is evident. Such occurrences' are common in most underground fires. Selfheating within the seam can serve as an ignition source for bush fires, forest fires [6] Further problems associated with such fires have been discussed extensively in literature [3, 7, 8], however there has been limited discussion on methodologies used for suppressing or controlling these fires.

Fig 1: Artistic conception of a smoldering fire in an abandoned coal mine and illustration of possible fire damage and suppression attempts (illustration from Hadden and Rein [5])

This paper discusses the option using CO_2 or a combination of CO_2-water vapor as a suppressing agent to extinguish or control these fires. Both are by-products of flue gasses from industrial power plants and thus can be produced abundantly if appropriate technology is implemented. The technique of using CO_2 as an extinguishing agent is definitely not new, with the earliest recorded instance of its use by Walker [9] in 1851. However, there has been limited work in this area from a fundamental viewpoint. Prior work has mostly focused on reporting case studies which are mine specific [9-14]. This paper discusses the relevant parameters that must be taken into account when dealing with suppression of such fires. The current state of knowledge and research areas that must be addressed before this technology can be applied in the field is discussed.

Parameters Involved in the study of Underground Coal Fire Suppression

The earliest use of CO_2 in suppressing an underground coal mine fire was reported by Walker [9] where he discusses a case by the Carbonic Acid Gas Co. of London, coal mine at Clackmannan, about 7 miles from Stirling in Scotland. There had been an underground fire in this mine, in the 9 foot deep coal, extending over an area of 26 acres, for 30 years. The plan adopted was to withdraw the men from the mine, and close all entrances to it, except two, one for the entrance, and the other for the escape of the gas. Carbon dioxide was made by forcing air through a coke furnace, the CO_2, N_2 and SO_2 so formed, in conjunction with a spray of water, was forced into the intake by the aid of a jet of steam. The process was maintained night and day for 3 weeks, the mixed gases and water being forced into the mine, during the whole of that period, and through the burning mass, and allowed to escape by the outlet mentioned. Cold air mixed with water spray was blown through the mass for a further period, and it is stated that at the end of a month, the fire was completely extinguished. Other studies of successful usage of CO_2 in coal mine fire suppression have been discussed by Morris [12] , Ray et al. [14]. and Sinha et al. [15]. The case study by Walker is interesting because it highlights the important parameters that need to be quantified before implementing CO_2 suppression on a larger scale. These parameters are listed in figure 2 and explained further.

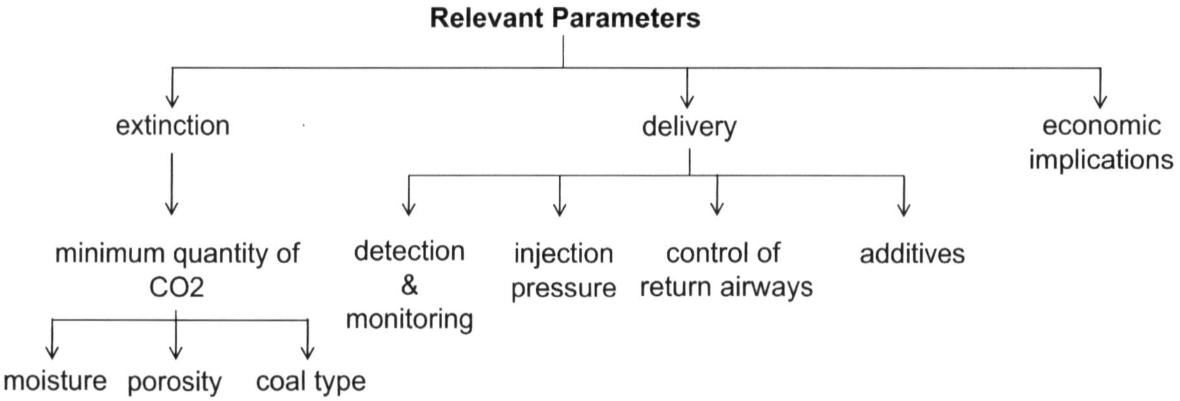

Figure 2: Flow chart representing the relevant parameters that need to be studied to implement CO_2 based control/suppression of underground fires

To simplify the problem, three areas need to be addressed:

1. Extinction and control:
   This refers to the analysis of the minimum quantity of suppressant needed to do the job. Both the quantity and application time need to be analyzed.
2. Delivery mechanism:
   The delivery of the suppressing agent to the seat of the fire is an important problem. Besides determining the optimum injection pressure, there needs to be some study of detecting and monitoring the growth/suppression of the fire. Secondly, additional additives such as water vapor, N2 that would aid in suppression also need to studied. It is also necessary to ensure that the return airways can be modeled.
3. Cost:
   The cost of installing a suppression system will be an important constraint. Ideally, a thorough cost benefit analysis of extinguishment of the fire with respect to the impact on the environment and local community needs to be analyzed. The costs would change based on the availability of CO_2, size of the fire, geography etc.

Each of these areas is further discussed in sections below:

Extinction/Control

Since oxygen has the strongest effect on a smoldering reaction, oxygen displacement is a highly effective method of extinguishing a coal fire. However, the quantity of inert gas required to displace sufficient oxygen needs to be quantified. Further this quantity will depend on the thermophysical properties of the coal, temperature, moisture and porosity. Figure 3 shows a hypothetical plot that shows the influence of CO_2 and autoignition temperature of coal. At a certain CO_2 concentration shown by the dashed line, autoignition is not possible. This zone is shaded and labeled “extinction region.”

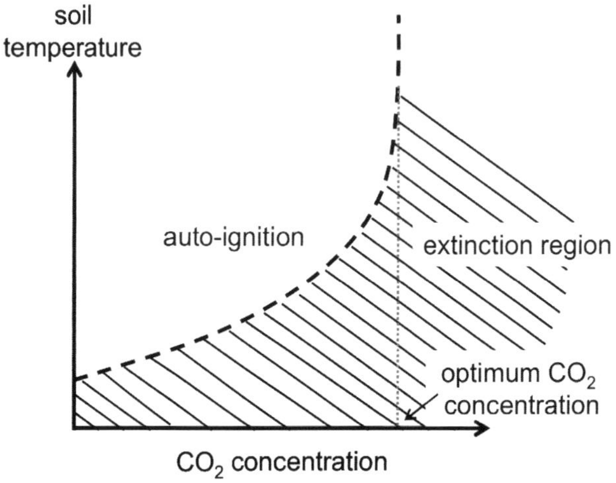

Figure 3: Functional relationship between autoignition temperature of coal and CO_2 concentration (hypothetical)

A functional relationship between minimum quantity of CO_2 and temperature therefore exists and must be determined to set the bounds for the quantity of CO_2 required to extinguish a particular seam of coal. An optimum concentration of CO_2 can then be obtained a plot as shown in figure 3. The authors are not aware of any standard test protocol capable of generating a plot such as figure 3. Further, coal seams extend several km with a thickness in the range of 1–6 m. The vast thickness of strata of low conductivity can maintain temperatures for a long period of time. This causes the inert environment created by CO_2 to merely suspend the smoldering combustion. As soon as CO_2 is stopped autoignition can occur relatively easily due to the trapped heat. Thus, suppression with CO_2 has to be coupled with some kind of monitoring system for temperature distributions in the strata. Typically, smoldering temperatures of coal are in the range of 750 – 950°C [16].with minimum self ignition temperatures varying between 50 – 150°C [17] depending on layer thickness, coal type and thermal conductivity of the strata. To aid cooling, water vapor can be added in addition to CO_2. Water vapor is a good agent to add to CO_2 due to its high specific heat.

Another method is to use liquid CO_2, which would act as a cooling agent as well as an inerting agent. In the US, liquid CO_2 has been used in the Bituner Mine. The quantity used was reported as 18,406 m^3, with an injection rate of 190 m^3/hr [13]. There are some instances where solid CO_2 (dry ice) has been used [18], by placing it directly on the fire scene. Use of dry ice would be useful to suppress smoldering fires in shallow fronts, where the ignition front is open to the atmosphere and having large supplies of oxygen available. In such cases, cooling the coal to a temperature below its critical autoignition temperature would be a more prudent recourse compared to displacement of oxygen.

For deep seated underground fires, oxygen transport is usually the limiting variable and thus CO_2 injection in the form of gas or liquid shows promise in such cases. Since CO_2 has a higher density compared to air, it is a good agent for fire located in depressions compared to other inert agents such ad N_2. If steam is added to CO_2, additional cooling can also be achieved due to the high specific heat of steam. The disadvantage of using CO_2 is the probability of CO production at high temperatures (~1000 – 1200 K) based on the Boudard reaction where CO_2 is converted to CO while consuming carbon $CO_2 + C \rightleftharpoons 2CO$ [17]. The CO produced could then react with steam leading to the water-gas reaction ($CO + H_2O \rightarrow CO_2 + H_2$) which would now promote rather than inhibit the reaction. Thus although CO_2 shows promise of being a cheap and effective extinguishing agent, there is a great need for further research to better understand the extinction mechanisms.

Delivery/Monitoring

Detection is the first step since it is impossible to suppress a coal fire without locating it first. Recent work by Kim [19] from the DOE National Energy Technology Laboratory (NETL) laboratory discusses methodologies that have been used in detecting underground coal fires. Use of satellite imagery has also been implemented to detect and monitor underground coal fires[20]

Cost/Benefit

The costs of containing, controlling, and extinguishing a fire can be so high that the only course of action, particularly for underground fires, is to let the fire burn itself out—this is the case for numerous fires in Pennsylvania (USA). However considering the impact in terms of green house gas emissions alone, this may not be a good strategy of control in today's regulatory environment. Agreed, there will be a cost associated with research on developing an effective suppression methodology, however, it is worthwhile to invest in this area. The need to reduce CO_2 emissions is an obvious motivating factor. A recent study by Keefe et al. [21] shows that CO_2 and Hg emissions from coal-fires in the U.S. are estimated at 1.4×10^7–2.9×10^8 t/yr and 0.58–11.5 t/yr, respectively. These estimated carbon dioxide emissions from underground coal fires represent roughly 1 – 10% of the annual emissions from coal-burning power plants. The world-wide carbon dioxide emissions from coal fires are unknown. The idea of using flue gasses rich in CO_2 from a power plant to extinguish a coal fire therefore deserves some attention in today's times, where environmental concerns are high. It is possible that some of the CO_2 that is utilized will be trapped in the coal bed, thus offering an efficient means of CO_2 sequestration as well. The cost of delivery and additives also need to be factored into the analysis of utilization of carbon dioxide for suppression and control of coal fires.

Conclusions

A brief literature review on the topic of underground coal fire suppression using CO_2 shows that there are several fundamental areas of research that need to be addressed before this idea can be put into practice. However, rapid industrialization, especially in growing economies such as China and India, coupled with the problem of global warming, should be reason enough to invest in this area of research. The ultimate application to a realistic fire requires an analysis of extinction kinetics, several transport issues related to the injection of CO_2 in a coal mine and the migration of the gas to the site of the fire.

References

1. Heffern, E.L. and D.A. Coates, *Geologic history of natural coal-bed fires, Powder River basin, USA, coal fires burning around the world: A global catastrophe.* International Journal of Coal Geology, Stracher, G.B., ed, 2004. **59**(1-2): p. 25-47.
2. Bell, F.G., S.E.T. Bullock, T.E.J. Hälbich, and P. Lindsay, *Environmental impacts associated with an abandoned mine in the Witbank coalfield, South Africa.* International Journal of Coal Geology, 2001. **45**: p. 195-216.
3. Nelson, M.I. and X.D. Chen, *Survey of experimental work on the self-heating and spontaneous combustion of coal.* The Geological Society of America: Reviews in Engineering Geology, 2007. **XVIII**: p. 31-83.
4. Zhang, X., J.L. van Genderen, H. Guan, and S. Kroonenberg, *Spatial analysis of thermal anomalies from airborne multi-spectral data.* International Journal of Remote Sensing, 2003. **24**(19): p. 3727-3742.
5. Hadden, R. and G. Rein, *Small scale extinguishing experiments of coal subsurface fires.* 6th Mediterranean Combustion Symposium, (2009)
6. Bustin, R.M. and W.H. Mathews, *In situ gasification of coal, a natural example: History, petrology, and mechanics of combustion.* Canadian Journal of Earth Sciences, 1982. **19**(3): p. 514-523.
7. Stracher, G.B. and T.P. Taylor, *Coal fires burning out of control around the world: Thermodynamic recipe for environmental catastrophe.* International Journal of Coal Geology, 2004. **59**(1-2): p. 7-17.
8. Krajick, K., *Fire in the hole-paging in mines from Pennsylvania to China, coal fires threaten towns, poison air and water, and add to global warming.* Smithsonian, 2005. **36**(2).
9. Walker, S.F., *The use of carbon dioxide.* Mines and Minerals, June Issue, 1908.
10. Greuer, R.E., *Study of mine fire fighting using inert gasses.* USBM contract report No. 50231075, Bureau of Mines Open File Report, 1974: p. 15-76.
11. Westfield, J., H.S. Brumbaugh, and R.W. Whittaker, *Extinguishing fire with carbon dioxide in the Valier mines.* USBM Information Circular 7563, 1950.

12. Morris, R., *A review of experiences on the use of inert gasses in mine fires.* Mining Sci. Tech., 1987. **6**: p. 37-69.
13. Ray, S.K., A. Zutshi, B.C. Bhowmick, N. Sahay, and R.P. Singh, *Fighting mine fires using gasses with particular reference to nitrogen.* J. South African Inst. Mining and Metallurgy, 2000: p. 265-272.
14. Ray, S.K. and R.P. Singh, *Recent developments and practices to control fire in underground coal mines.* Fire Tech., 2007. **43**: p. 285-300.
15. Sinha, P.R., *Mine fires in Indian coalfields.* Energy, 1986. **11**(11-12): p. 1147-1154.
16. Rein, G., S. Cohen, and A. Simeoni, *Carbon emissions from smoldering peat in shallow and strong fronts.* Proceedings of the Combustion Institute, 2009. **32**(2): p. 2489-2496.
17. Haghighi, M., Z. Sun, J. Wu, J. Bromly, H.L. Wee, E. Ng, Y. Wang, and D. Zhang, *On the reaction mechanism of CO_2 reforming of methane over a bed of coal char.* Proceedings of the Combustion Institute, 2007. **31**(2): p. 1983-1990.
18. Bacharach, J.P.L., A.L. Craven, and D.B. Stewart, *Underground mine fire control with inerting systems.* CIM Bulletin, 1986. **79**: p. 67-72.
19. Kim, A.G., *Locating fires in abandoned underground coal mines.* International Journal of Coal Geology, 2004. **59**(1-2): p. 49-62.
20. Voigt, S., A. Tetzlaff, J. Zhang, C. Künzer, B. Zhukov, G. Strunz, D. Oertel, A. Roth, P. van Dijk, and H. Mehl, *Integrating satellite remote sensing techniques for detection and analysis of uncontrolled coal seam fires in north China.* International Journal of Coal Geology, 2004. **59**(1-2): p. 121-136.
21. O'Keefe, J.M.K., K.R. Henke, J.C. Hower, M.A. Engle, G.B. Stracher, J.D. Stucker, J.W. Drew, W.D. Staggs, T.M. Murray, and M.L. Hammond Iii, *Co2, co, and hg emissions from the Truman Shepherd and Ruth Mullins coal fires, eastern Kentucky, USA.* Science of The Total Environment, 2010.

Possible Use for CO_2 Captured from Coal Derived Flue Gas: Reforming of Methane in Electric Discharge Plasmas

Erik C. Rupp, ORISE Postdoctoral Researcher, National Energy Technology Laboratory, Pittsburgh, PA
Evan J. Granite, National Energy Technology Laboratory, Pittsburgh, PA

Coal-derived energy is perhaps the most important, and accessible, source of energy in the world. According to the United States Energy Information Administration, the US consumed over 1 billion metric tons of coal in 2006, with the entire world consuming over 6.5 billion metric tons. During this same year, the world consumed 3 trillion m^3 of natural gas, with the US contributing 600 billion m^3 to this total. This resulted in the emission in the US of 2 billion metric tons of CO_2 from coal and 1 billion metric tons of CO_2 from natural gas, or just over half the total CO_2 emissions from all sources in the United States during 2006. As the worlds' understanding of anthropogenic climate change increases, it becomes clear that CO_2 emissions must be abated, with CO_2 formed from stationary sources such as power plants potential the easiest to capture or convert to useful products.

The current direction for the use of CO_2 involves sequestration in geological formations after capture from flue gas. The US Department of Energy's Fossil Fuel Program's carbon sequestration core and the FutureGen Initiative are major spearheads for this technology. Other proposals for the use of the captured CO_2 involve advanced catalysts, engineered photosynthesis systems, biomimetic systems and biological systems. The use of CO_2, natural gas and electrical plasma discharges could provide another option for the use of captured CO_2, through the production of syngas, which is a useful product for either energy generation or in the Fischer-Tropsch process to produce synthetic liquid hydrocarbons.

The interaction of plasma discharges with CO_2, and the CO_2 – CO equilibrium in the plasma has been studied since the late 19th century with research picking up in the early 20th century (Wendt, Hunt). This work typically focused on corona discharges and equilibrium conditions. This work was so early that it tended to be a series of observations without attempts at explanation. In the middle of the 20th century, it was recognized that energetic electrons were the likely cause of CO_2 conversion and equilibrium with CO (Mizushima). It was eventually understood that CO_2 – CO equilibrium in a plasma discharge involved complex chemistry, involving the surfaces of the electrodes (Barton) and the input power (Brown). Although there is theoretical importance to this work, the equilibrium of CO_2 and CO has little practical value as the conversion of CO_2 is typically low. More recent work has focused on the use of CO_2 with CH_4 in a variety of different plasma discharges.

In general, there are two categories of plasma discharges, thermal and non-thermal. The major difference between the two is whether the gas stream is heated or if only the electrons are at a higher temperature, with the gas stream unheated. In non-thermal plasmas, the majority of the electrical energy provided to the gas stream is used to produce energetic electrons, with little heating of the gas stream. There is also a difference in operating pressures, as thermal plasma discharges tend to operate at pressures, while non-isothermal plasma discharges can operate at atmospheric or near-atmospheric pressures.

The differences in the two operation categories allow for different uses and create specific advantages and disadvantages. Thermal plasma reactors, typically created by arcs or radiofrequency (RF) inductively coupled discharge, provide high power output at high pressures. However, they are limited by low energy efficiencies due to low selectivity, high gas temperatures and electrode design issues. There is a variety of non-thermal plasma discharges, including low pressure glow, corona, microwave, dielectric barrier, gliding arc and spark discharge. Fridman et al. reviewed several of these non-thermal discharges, discussing the physics and primary features of the discharges. A number of other sources provide key background information for plasma discharges(van Veldhuizen, Lieberman, Razier).

This review is intended to discuss the state of technology for CO_2 reforming of CH_4 in plasma discharges. A number of discharge technologies (corona discharge, dielectric barrier or silent discharge, radio frequency discharge, thermal arc discharge) will be presented, with particular emphasis placed on CO and H_2 selectivity, removal of CO_2 and CH_4 and energy efficiency. A discussion about the viability of the range of technologies for the use of CO_2 and CH_4 to produce syngas, as well as key technological hurdles, concludes the review.

References

H. Hunt and W.C. Schumb, *J. Am. Chem. Soc.*, 1930, **52**, p. 3152
G. Wendt and M. Farnsworth, *J. Am. Chem. Soc.*, 1925, **47**, p. 2494
S. Mizushima, Y. Morino, H.H. Hirabayashi, K. Kakihara and Y. Mizushima, *J. Am. Chem. Soc.*, 1950, **72**, 51, p. 5176
S.S. Barton, S.T. Balke, R.J. Beardmore, R.T. Perry, M. Daniel and W. Newstead, *J. Phys. Chem.*, 1967, **71**, p. 4573
L. Brown and T. Bell, *Ind. Chem. Eng. Fund.*, 1974, **13**, p. 203 and p. 210
E.M. van Veldhuizen, *Electrical Discharges for Environmental Purposes: Fundamentals and Applications*, 2000, Nova Science Publishers; Huntington, NY
M.A. Lieberman and A.J. Lichtenberg, *Principles of Plasma Discharges and Materials Processing,* 1994, Wiley; New York
Y.P. Razier, *Gas Discharge Physics*, 1991, Springer-Verlag; New York
A. Fridman, A. Chirkov and A. Gutsol, *J. Phys. D: Appl. Phys.*, 2005, **38**, p. R1

(157c) Survey of Potential Uses for Carbon Dioxide Captured From Coal-Burning Power Plants

Evan J. Granite, National Energy Technology Laboratory, Pittsburgh, PA

Erik C. Rupp, National Energy Technology Laboratory, Pittsburgh, PA

Bryan D. Morreale, National Energy Technology Laboratory, Pittsburgh, PA

Henry W. Pennline, National Energy Technology Laboratory, Pittsburgh, PA

Dirk Link, National Energy Technology Laboratory, Pittsburgh, PA

David Luebke, National Energy Technology Laboratory, Pittsburgh, PA

Todd H. Gardner, National Energy Technology Laboratory, Morgantown, WV

Approximately 50% of the electricity in the United States is generated through the combustion of coal. One billion tons of coal is burned annually for electricity production, resulting in over two billion tons of carbon dioxide. Carbon dioxide is a potent greenhouse gas that can contribute to global warming. The United States Department of Energy is sponsoring research on the separation and sequestration of carbon dioxide from coal-derived flue gas. A major issue is determining good uses for the massive quantities of carbon dioxide that could potentially be separated from flue gas.

Some of the current uses of carbon dioxide that will be examined include fire suppression, food preservation, refrigeration and cooling, beverage carbonation, inert gas needs such as for welding, enhanced oil and coal bed methane recovery, supercritical cleaning, biomass production, and polymer-plastic manufacture. Future research areas for carbon dioxide utilization such as fuel production, pharmaceutical chemicals, coal gasification, oxy-combustion of coal, and passivation of alkaline wastes will also be discussed.

Literature Cited

1. Edwards, J.H. Potential sources of CO2 and the options for its large-scale utilization now and in the future, Catalysis Today, 23, 59-66, 1995.

2. White, C.M.; Strazisar, B.R.; Granite, E.J.; Hoffman, J.S.; Pennline, H.W. Separation and Capture of CO2 from large Stationary Sources and Sequestration in Geological Formations – Coalbeds and Deep Saline Aquifers, Journal of the Air & Waste Management Association, 53, 645-715, 2003.

See more of this Session: Beneficial Utilization of Carbon Dioxide Separated From Flue Gas
See more of this Group/Topical: Topical 2: Advanced Fossil Energy Utilization

(159a) Study On Flare Minimization During Ethylene Plant Upsets

Xiongtao Yang, Dan F. Smith Department of Chemical Engineering, Lamar University, Beaumont, TX

Qiang Xu, Dan F. Smith Department of Chemical Engineering, Lamar University, Beaumont, TX

Kuyen Li, Dan F. Smith Department of Chemical Engineering, Lamar University, Beaumont, TX

Process upsets mean facilities run in an unstable or an undesirable situation that may jeopardize product quality and/or plant safety. During the ethylene production, plants might experience various process upsets. Some are significant and others are minor, and the majority contributions to flaring emission are from acetylene reactor effluent and ethylene splitter. To reduce the flaring caused by process upsets, recycling off-spec materials to the upstream process for reprocessing presents a good choice. Two major options are available to routine the off-spec components: one is to line the recycle to the CGC suction, and the other is to tie it to the furnace feed. Establishing these two recycles presents a promising way and brings a lot of benefits, which include reducing flaring emission, reducing fresh feedstock, increasing the tolerance of the process when facing upsets, avoiding unnecessary plant shutdowns and subsequent startups, saving energy and fuel, etc.

This paper conducts quantitative study to help plants minimize flaring without sacrificing safety during process upsets. As the two recycles have their limits, they should be implemented carefully. Considering their limits and safety issues, various upset scenarios are extensively analyzed. Dynamic simulation-based solutions are provided accordingly, such that efficient and effective ways can be taken to reduce upset related flaring.

See more of this Session: Innovations in Environmental Emission Abatement
See more of this Group/Topical: Process Development Division

(159b) Simulating Industrial Ethylene Flares

Hitesh Vaid, Chemical, Lamar University, Beaumont, TX

Anjan Tula, Chemical, Lamar University, Beaumont, TX

Kanwar Devesh Singh, Chemical, Lamar University, Beaumont, TX

Daniel H. Chen, Chemical Engineering, Lamar University, Beaumont, TX

Helen H. Lou, Department of Chemical Engineering, Lamar University, Beaumont, TX

Kuyen Li, Lamar University, Beaumont, TX

Xianchang Li, Mechanical Engineering, Lamar University, Beaumont, TX

Christopher Martin, Chemical, Lamar University, Beaumont, TX

Air quality models (CAMx, CMAQ) often significantly under-predict observed peak O3 with the current VOC emission inventories in HGB/BPA. Even though this could be due to the inaccuracy of the photochemical models used, it is also likely that some VOC emissions may be under-reported or unreported due to the assumption of robust destruction efficiencies of industrial flares. According to TCEQ's 2007 HRVOC special inventory, industrial flares are a major source of highly-reactive VOCs (HRVOCs). From rigorous combustion chemistry, incomplete combustion of hydrocarbons is a likely source of free radicals such as perhydroxyl and radical producing species such as formaldehyde. All together, these pollutants may be responsible for the particularly high ozone formation in the Houston-Galveston-Brazoria area. Our simulations of industrial-sized ethylene flares based on FLUENT and combined San Diego /GRI mechanisms show the carbon-based flare efficiency (CBFE) decreases with cross wind speed even though the destruction and removal efficiency (DRE) stays relatively constant. Over-aeration also reduces the average flame temperature, therefore increases the incomplete-combustion-product emissions. The studies are aimed at finding the relationship between flare efficiencies/emissions and flare operating/meteorological conditions. Another important goal is to help petroleum/chemical industries to establish flare operation guidelines if enough simulation data are generated and validated with available experimental flare/flame data.

See more of this Session: Innovations in Environmental Emission Abatement

See more of this Group/Topical: Process Development Division

(159c) Proactive Flare Minimization During Chemical Plant Startups

Chaowei Liu, Dan F. Smith Department of Chemical Engineering, Lamar University, Beaumont, TX

Xiongtao Yang, Dan F. Smith Department of Chemical Engineering, Lamar University, Beaumont, TX

Qiang Xu, Dan F. Smith Department of Chemical Engineering, Lamar University, Beaumont, TX

Kuyen Li, Dan F. Smith Department of Chemical Engineering, Lamar University, Beaumont, TX

Flaring emission during chemical plant turnaround operations generates huge amounts of off-spec products for flaring, and thus causes negative environmental and social impacts and also results in tremendous raw material and energy losses. For instance, an ethylene plant with 1.2 billion pounds of ethylene production per year may flare five million pounds of ethylene during one typical startup, which generates 15.4 million pounds of CO_2, 7.5 K lbs of NOX, 40 K lbs of CO, 100 K lbs of highly reactive VOCs. Obviously, minimizing flaring has great environmental, societal, and economic significance.

Based on previous studies on flare minimization at Lamar University, this work will address the proactive flare emission reduction strategy. In this strategy, process design and operation will be integrated together to reduce the startup time, so as to reduce the amount of flaring sources. The simulated dynamic emission profiles for various species during a plant startup will be presented. The potential environmental impacts and economic benefits derived from the proposed flare minimization solutions are also provided.

See more of this Session: Innovations in Environmental Emission Abatement
See more of this Group/Topical: Process Development Division

(159d) Modeling and Optimization of a Packed-Bed Tubular Reactor for Eco-Friendly Epichlorohydrin Production

Woohyun Kim, Department of Chemical & Biomolecular Engineering, Korea Advanced Institute of Science and Technology, Daejeon, South Korea

Choamun Yun, Department of Chemical & Biomolecular Engineering, Korea Advanced Institute of Science and Technology, Daejeon, South Korea

Young Kim, Korea Institute of Machinery and Materials, Daejeon, South Korea

Jeongho Park, Department of Chemical & Biomolecular Engineering, Korea Advanced Institute of Science and Technology, Daejeon, South Korea

Sunwon Park, Department of Chemical & Biomolecular Engineering, Korea Advanced Institute of Science and Technology, Daejeon, South Korea

Ki Taek Jung, Hanwha Chemical Corporation R&D Center, Daejeon, South Korea

Kyoung Sup Woo, Hanwha Chemical Corporation R&D Center, Daejeon, South Korea

Yong Hwa Lee, Hanwha Chemical Corporation R&D Center, Daejeon, South Korea

Sae Heon Kim, Hanwha Chemical Corporation R&D Center, Daejeon, South Korea

A liquid-phase epoxidation process using hydrogen peroxide (HP) and titanium silicate (TS-1) is an eco-friendly process to produce epichlorohydrin because high conversion and selectivity can be obtained over the catalyst with reduced waste water discharge. In this work, a mathematical model was built to predict catalyst deactivation, temperature profile, as well as time-varying conversion and selectivity of a packed-bed tubular reactor using HP and TS-1. The first step toward this work is the kinetic modeling of this catalytic reaction which consists of reaction mechanism selection and kinetic parameter estimation. Based on published literatures on the titanium-silicate catalyzed epoxidation of olefins with HP, the reaction mechanism was selected as the Eley-Rideal mechanism and the kinetic parameters were estimated by a hybrid optimization technique combining the Levenberg-Marquardt algorithm and genetic algorithms. In addition to the kinetic modeling of the catalytic reaction, the characteristics of a commercial TS-1 catalyst pellet, such as time-varying catalytic activity and effectiveness factor, were also considered. To build the mathematical model of the pilot-scale reactor, the kinetic model and the characteristics of the TS-1 catalyst pellet were included in 2-dimensional dynamic mass and energy transport models for a packed bed tubular reactor. Therefore, the time-varying temperature and concentration profiles of the reactor were obtained by dynamic simulations. Based on the simulation results, the relationships between operating conditions and the conversion and selectivity of the tubular reactor are analyzed. To maximize catalyst activity in a long-term operation, the optimal operating temperature can be identified by controlling the hot spots of the reactor. Also, the optimal length of the reactor is determined to maintain desired conversion and selectivity. Simulation results show that the optimally-sized reactor operating at the optimal conditions may maintain both conversion and selectivity above 96% during 2,000 hours.

Acknowledgement: This work was supported by Hanwha Chemical R&D Center.

See more of this Session: Innovations in Environmental Emission Abatement
See more of this Group/Topical: Process Development Division

Synthesis of α-hydroxy dodecanoic acid by hydrolysis of α-chlorine dodecanoic acid

Wenfu CHEN, Yun FANG*

School of Chemical and Material Engineering, Jiangnan University, Wuxi, Jiangsu 214122, China

The long-chain α-hydroxy fatty acid is one of important fine chemicals, which can be widely used in flavors, fragrances and medicines [1]. The traditional methods to synthesize the long-chain α-hydroxy fatty acid usually employed alkali as the hydrolytic reagents [2], which bearing many drawbacks such as environmental pollution and impurities in the final products.

In this paper, we reported a novel preparation procedure of α-hydroxy dodecanoic acid (α-HDA) by hydrolysis of α-chlorine dodecanoic acid (α-CDA) using calcium carbonate. In this procedure, $CaCO_3$ can be recycled and reused, which would decrease the discharge amounts of waste alkali solutions effectively. The preparation procedure of α-HDA is shown as following.

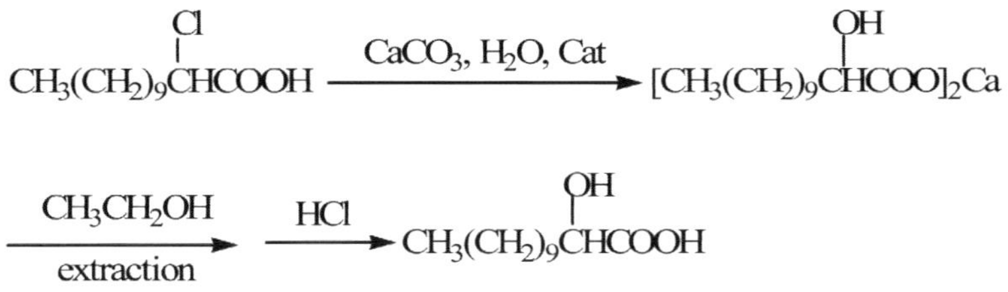

The effects of added water, reaction temperature, starting materials ratio, and reaction time were investigated carefully. Under the optimized condition, a yield of 95.2% of α-HDA would be obtained after 6 hours reaction at 130 °C with n(CaCO3):n(α-CDA)=3.5, m(Cat):m(α-CDA)=0.02, and m(H2O):m(α-CDA)=2. The purity of α-HDA is 99.6% after recrystallization in petroleum ether.

References

1 Yu Ruey, Van Scott Eugene. Pharmaceutical compositions containing hydroxycarboxylic acids and/or ketocarboxylic acids and methods of using same. US 5962526, 1999.

2 Le Sueur, Henry Rondel.The effect of heat on α-hydroxy carboxylic acids. *Journal of the Chemical Society, Transactions.* 1905; 87: 1888-1906.

* Corresponding author: yunfang@126.com. Supported by the National Natural Science Foundation of China (20871059, 20371021).

AUTHOR INDEX

AUTHOR INDEX

American Institute of Chemical Engineers
3 Park Avenue
New York, New York 10016-5991

proceedings.com